Amber

Amber

From Antiquity to Eternity

Rachel King

REAKTION BOOKS

For Iris, for Benoît, for Thilo

Published by Reaktion Books Ltd
Unit 32, Waterside
44–48 Wharf Road
London N1 7UX, UK
www.reaktionbooks.co.uk

First published 2022
Copyright © Rachel King 2022

Printed and bound in India by Replika Press Pvt. Ltd

A catalogue record for this book is available from the British Library

ISBN 978 1 78914 591 5

Contents

moment the resin is extruded and meets the light. It continues as the resin matures, usually buried by soil or sediments where it is affected by the different elements, temperatures and pressure of the local environment, as well as the oxygen content thereof. Amberization, as this process of maturation is sometimes called, ends when the substance has lost all of its volatile constituents and become completely stable. Confusingly, both the resin and the plant and animal remains it entombs are referred to as fossils.

Young, recently hardened resins are called copals. The word 'copal' derives from the Nahuatl word *copalli*, and, like amber, has a complicated history. In pre-Columbian Meso-America, the word had a locally specific meaning. Over the last five hundred years, the word has expanded to mean a variety of subfossil resins. In copals, the

4 Reconstruction of the possible appearance of the ancient amber forest. Note the resin seeping from the trees and dripping to the ground and into the water. Trees, parts of them, and drops of resin may also have been carried away and deposited downstream.

molecules are incompletely cross-linked and some are still volatile. A drop of alcohol makes their surfaces tacky; a flame melts them. Copals can be polished, but they will quickly craze and crack as the freshly exposed oils, acids, alcohols and volatile aromatic components evaporate. These components will gradually be lost if copals are left in the soil and the conditions are right. There are, in effect, many ambers because everything along the sliding scale of hardness can be, or may at one time or other have been, perceived to be amber. Culturally and historically speaking, amber is the name for a wide variety of scientifically diverse yellow-looking resins.

No single feature determines when exactly copals mature into amber. Age naturally increases maturity, but the rate at which maturity proceeds depends on the original constituent components of the resin and on geological conditions. These conditions are not always consistent across time. Though some ambers are found in situ, many are found in localities far from the actual place where the resin was extruded. Some deposits are thought to have been amassed when droplets of resin were washed into streams and transported. They were dropped where the water was shallow and slow-flowing, in deltas and lagoons, and buried by sediments and silts. It is also thought that tree branches and trunks were sometimes transported by water, and resin drops adhering to them were similarly transformed (illus. 4).[3] In another scenario, already-fossilized amber is thought to have been eroded from its original setting and retransported by glacial processes.

SUCCINITE – BALTIC AMBER

For most Europeans, particularly northern Europeans, amber has historically been the name for a fossil resin from the Baltic region. This is home to one of the world's largest deposits of amber, something about the size of Wales. Baltic amber is also one of the longest-exploited and most scientifically, culturally and economically significant deposits. It is estimated to be the source of nine-tenths of the amber ever traded in Europe, and until the twentieth century,

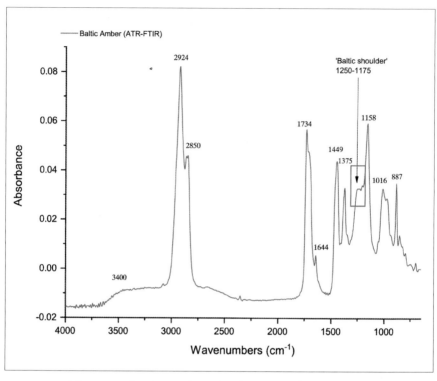

5 Reference FTIR (Fourier-transform infrared spectroscopy) spectrum of Baltic amber with the absorption peak known as the 'Baltic Shoulder' highlighted.

when Dominican amber and amber from Myanmar became more widely known, Baltic amber dominated the international market.

The deposit contains fossilized resins from several different sources. The principal fossilized resin found in the Baltic region has the mineralogical name succinite. Coined in the 1820s, this name comes from the Latin *succinum* (from *succus*, meaning 'juice'). It is specifically applied to amber containing between 3 and 8 per cent succinic acid. This represents about nine-tenths of all Baltic amber. When succinite is tested using infrared spectroscopy, the resulting spectrum shows a plateau on the side of one of the peaks called the 'Baltic Shoulder' (illus. 5).[4] However, about eighty further types of amber found in the same region contain little or no succinic acid.[5] These each have their own names: for example gedanite, named after the city of Gdańsk; stantienite and beckerite, named after the mining firm Stantien & Becker; and glessite, after the Latin term *glaesum*, said by the Roman historian Tacitus to have been the local name for amber.

HOW OLD IS AMBER?

For all its fame, Baltic-origin amber is *not* the oldest in the world. It is made from resin formed between 48 and 34 million years ago (MYA). The oldest fossil resin is found only in microscopic trace quantities and is nearly seven times older. It was formed in about 320 MYA. This early amber is referred to as Carboniferous, after the name of the period in which it was formed.

Because the limit for carbon-14 techniques is set at 40,000 years, it is impossible to test amber itself for age. Contextual stratigraphic data therefore supplies significant information for those who wish to date ambers. In many cases, however, the geological and geographical origins of historically collected older specimens are not known. Foreign bodies and particles can be useful clues. Organisms such as bacteria and protozoa, as well as fungal and plant spores are found in amber of the Triassic period, about 230 MYA, while many ambers were formed between about 146 and 66 MYA, an era known as the Cretaceous period. Flowering plants (angiosperms) are not known before the Cretaceous, although terrestrial conifers and cycads (palm-like cone-bearing trees) are. Ambers from the Cretaceous and Tertiary (66–2.6 MYA) periods have been found in North America, the Dominican Republic, Ethiopia, France, India, Israel, Japan, Mexico, New Zealand, Siberia, Spain and Switzerland. Japanese amber in particular has been found to hold fossilized marine life, insects and feathers. In recent years, amber from Myanmar, known as burmite, has also become more available and subsequently better understood. Palaeontologists have been able to use the many and varied inclusions it embalms, as well as radiometric dating of soil and ash samples, to re-date the Hukawng Valley deposit to circa 99 million years old. Burmese and Baltic amber were previously considered contemporaneous.

CAUGHT IN THE ACT

Few gemstones hit the headlines as often and as sensationally as amber. Recent reports have focused on Myanmar and celebrated the sensational discovery of a '100 million-year-old dinosaur tail', discussed 'ancient birds' wings preserved in amber', explored the 'lost world' revealed by amber-trapped lizard fossils and examined 'extinct plant species discovered'.[6] The words journalists choose – ancient, reveal, lost and extinct – capture and communicate the popular fascination for amber's mysteries. Book titles fete the material as a magical natural time capsule, and down the ages writers have lamented that the creatures it encloses cannot tell them how they became entrapped.

This fascination is entirely justified. Amber is unique in its ability to preserve delicate organisms and entities. From almost invisible bacteria to squat frogs, bubbles of air and water, faeces, minerals and sometimes drips of previously exuded resin, amber is the ultimate and most indiscriminate preserver. The evidence found in amber is virtually unparalleled in the geological record. It preserves soft tissues that supply a stock of potential information about cells and body chemistry. Considered individually and as a whole, inclusions in amber are an extraordinary, often amazing, source of information about the history of life on earth.[7] This not only compels an intense global traffic and generates high prices; sometimes, it tragically drives illegal trade and with this human rights abuses, as well as cultural and environmental degradation.[8]

Over thirty deposits of insect-containing amber are known worldwide. These have produced the world's oldest known spider, feather-feeding insect, mushroom, tick, cobweb and pollinating bee. The amber they contain has preserved interactions which would otherwise be lost, such as mating insects, predators attacking prey, egg-laying, eating and ride-hitching. A single piece in Stuttgart has been found to entomb some two hundred individual insects, belonging to 22 different families.

Bustling worlds captured in amber remind us that resin was one element of an extensive ecosystem. Inclusions help understand the bigger picture, and not just because they too were once part of it.

For example, not everything has been preserved, and the presence of some insects is used to infer the existence of certain flora. Dominican amber shows such biodiversity that some scientists have proposed that the island's ancient forest ecosystem and climate could be re-constructed.[9] Others challenge this idea. They point out that some ambers were clearly formed under exceptional conditions: periods of abundant resin production (hyper-resinosis) for example. Some scientists now hold there to have been four significant 'amber bursts' in the geological record. What this means is still being debated. Were the forests dominated by trees steadily dripping resin? Or was a lot of resin produced by a small number of trees? Did the resin accumulate steadily over an extended period, or was it suddenly catastrophically released in massive amounts in response to stresses or to local or global occurrences?[10] The answers, should they ever be conclusively found, will significantly affect the ways in which the entombed fossil record is studied.

DINOSAURS AND DNA

Certain directions of research are already well established. For well over a century, people have also tried to extract biological information from insects embalmed in amber. One of the first successful attempts was made in 1903 by the Russian scientist Nicolai Kornilowitch when studying Baltic amber. Using no more than conventional light micro-scopy and histology, he found banding patterns typical of modern muscle fibres in tissue extracted from insects. In the early 1980s, scientists published images of the subcellular structures in tissue from a fungus gnat taken using an electron microscope – work then widely credited with opening the door for future research on the possibility of extracting ancient DNA.[11]

DNA and dinosaurs has become an obsession of popular science. Who does not know the 1993 film *Jurassic Park*, based on a novel by Michael Crichton of the same name? Its plot revolves around a group of scientists who bring dinosaurs back to life. They use DNA extracted from dinosaur blood in the stomachs of mosquitoes that had fed moments before amber engulfed them.[12] When the film was made there were neither examples of insects in Jurassic amber, nor examples of mosquitoes embalmed in amber exuded in the actual era of the dinosaurs. Since then, only very few mosquitoes have been found.[13] When, in 2017, a tick was found clinging to a dinosaur feather in amber dating back 99 million years, there was much excitement. But even after death, stomach acids continue to digest and the DNA had been broken down too much to be analysed.[14]

Scientists have also looked to extract and sequence DNA from inclusions themselves. Their colleagues working on the human genome can draw on an almost limitless supply of tissue and fully intact

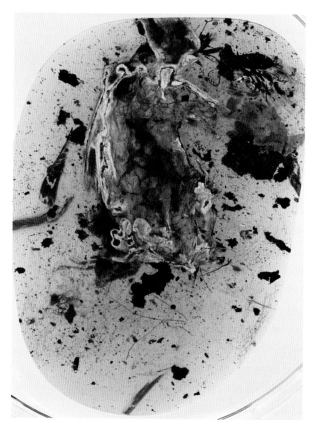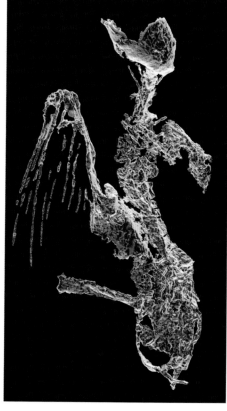

6, 7 Two images showing a dinosaur-era bird in amber: through the amber, peering into the bird's body cavity on the belly side (left), and a CT (computer tomography) reconstruction of the skeleton observed from the bird's back (right).

sequences of DNA. DNA preserved in amber fossils, on the other hand, is broken, disorderly and degraded. Trying to read and reassemble the genome of an extinct termite has been likened to trying to reconstruct a novel without ever having read the text and by fishing alphabet noodles out of soup. Thirty years ago, a new technique was used to extract, replicate and multiply fragmentary DNA from an extinct species of bee and from a termite encased in Dominican amber.[15] DNA has subsequently been recovered from fruit flies, wood and fungus gnats, lizards and leaf beetles, with approximately one in every three attempts meeting with success. However, much of this work now seems discredited and the positive results attributed to contamination.

Most recently the remains of birds have – sensationally – been found in burmite (illus. 6, 7), renewing speculation: if the science allowed it, could the bird be resurrected?[16] Even in the eminent case of the remarkably well-preserved, frozen 41,800-year-old woolly mammoth calf, discovered in Siberia in 2007, it was impossible to extract a complete cell, as is needed for cloning. If DNA cannot be obtained from bone and tissue that is less than 50,000 years old, is there any chance of obtaining good DNA from a piece of amber at 65 million years old?

No doubt many more attempts will be made to extract DNA. These will certainly have significant implications for the understanding of evolution, but this science is not all emphatically positive. There are ethical factors, especially if the culture of extinct microorganisms is the goal. In 1995, scientists reported that they had been able to culture a bacterium which they had extracted from a stingless bee preserved in amber. This bacterium is of particular interest because it is similar, but not identical, to another type of bacterium in modern bees.[17] If DNA is preserved, then the preservation of problematic viruses, bacteria, protozoa and fungi is also possible, raising safety concerns as well as issues relating to human responsibility towards the environment and museum collections held in trust for future generations. Obtaining amber through mining, extracting inclusions from within it and destroying specimens to obtain DNA are all irreversible processes.

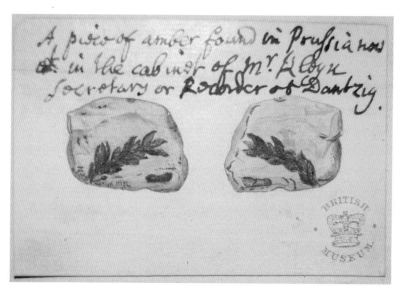

8 Mark Catesby, 'Two Studies of a Plant in Amber', 1727–49, ink and wash on paper, with an inscription by the English naturalist that reads: 'A piece of amber found in Prussia now in the cabinet of Mr. Kloyk, secretary or recorder of Dantzig.'

WHAT PRODUCED AMBER?

Lizards, frogs and flies in amber are delightful puzzles. Though they suggest something of the world in which amber was formed, they reveal little about the origins of the amber in which they are entombed. For centuries, people referred to enclosed leaves, petals, catkins, cones and mosses as circumstantial evidence that amber must be a tree resin. Classical authors in the West noted that burnt amber smelled of pine trees, and early modern European authors wrote that 'the manifest smell of pine left on fingers which have stroked it' or hanging in the air after burning it 'corroborate that amber [is] the gum of a pine'.[18] Collectors of specimens from the Baltic remarked that amber was shaped like icicles and drops, or sometimes bore the imprints of bark and leaves (illus. 8). In the 1880s, German botanist Heinrich Göppert gave the unknown mother-tree a name. He called it *Pinites succinifer*.[19] In Asia, where commentaries on the nature of amber go back at least 1,400 years, authors variously proposed that amber was fir sap or resin. The secretion penetrated the earth, congealed and hardened over a period of anywhere between a hundred years and a millennium.[20]

Why did it take so long for scientists to become sure that amber was the product of a tree? In the West, scholarship on amber has focused on the Baltic region. For centuries amber there was retrieved from the sea or dug from pits in the sand dunes, disguising any link to trees. The earliest surviving treatise to muse about this problem at length was written around 1538 by Gregor Duncker, a doctor who had lived at Fischhausen (present-day Primorsk) in southern Samland, now part of the Kaliningrad Oblast, for most of his life (illus. 9). He was reluctant to criticize ancient authors who claimed that amber was sylvan; but he also thought it unlikely that any had been to the Baltic. As a local, Duncker knew only of pines growing in Norway, from which he inferred that amber, if the result of pines, must have travelled impossibly far and been channelled through the narrow Sound strait at Copenhagen.

9 Fischhausen in the later 17th century, illustration from Christoph Hartknoch, *Alt- und Neues Preussen Oder Preussischer Historien Zwey Theile* (1684).

Duncker agreed that amber had started out as a liquid; he even wrote that he had seen it in its soft state.[21] But what sort of liquid? His rough contemporary Georg Bauer/Agricola felt that it could not be the juice of a tree, for sap was drawn out by the sun. He could not understand how there could be lots of amber in the sun-starved north, but none, as far as he was aware, in the tropics. Going on amber's appearance, flammability and range of colours, he proposed that it was a type of bituminous pitch.[22] Writing around the same time, Andreas Goldschmidt (pen name Andreas Aurifaber) concurred. Amber did not dissolve in hot water, so it could not be a resin. It burned, so it could not be a mineral. He agreed that it was created in the bowels of the earth and proposed its manner of finding as evidence enough of that. Some recent inland finds had been exceptionally large, and Goldschmidt argued this was because there were no waves on the ponds and lakes to disperse the initial gooey mass, as otherwise usually happened.[23] The debate was still raging one hundred years later in the 1660s. Fellows of the Royal Society in London wrote asking for clarification from the Danzig-based astronomer Johannes Hevelius and received the answer that amber was 'a kind of fossil pitch or bitumen'.[24] A century later, the famous Swedish botanist Carl Linnaeus was still discussing amber as 'born of pitch'.[25]

ANALYSING AMBER

German speakers have historically dominated the field of amber studies. Prussia was home to such extensive deposits of amber that local universities and scholars amassed high-quality specimen collections. German advances in microscopy placed its scientists at the forefront of many breakthroughs. Working at the end of the nineteenth century, modern equipment furnished botanist Hugo Conwentz with the tools to challenge previous generations (illus. 10). He had long studied the flora encased in Baltic amber.[26] It was thought that plant inclusions would be the key to the type of tree that made amber. He proposed the amendment of Göppert's *Pinites succinifer* to *Pinus succinifer*, a subtle but meaningful change. The suffix *-ites* implied that

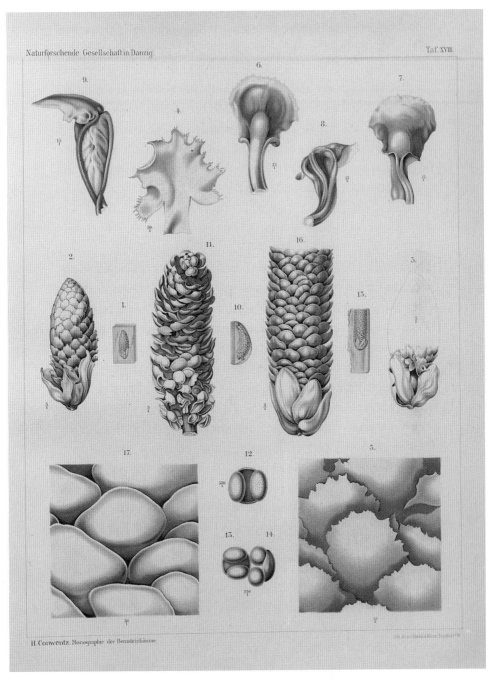

10 Plate showing flowers in amber, from Hugo Conwentz,
Monographie der baltischen Bernsteinbäume (1890).

amber originated with a single mother-tree and that this was ancient relative of the modern living species *Pinus*. Conwentz recognized that fragments of wood he was seeing were not ancient relatives of any modern pine tree and suggested *Pinus* be used as a term suitably encompassing the possibility that it was the resin of more than one pine variety and maybe even of spruce. His decision to use such a term as *Pinus* is responsible for much confusion and the misleading notion that a modern tree of the *Pinus* species was the source of the resin that fossilized into succinite.[27]

TODAY'S SCIENCE

Today nearly all scientists concur, based on analysis of amber's chemistry, that it did not come from a tree of the modern genus *Pinus*. They now recognize that the flora trapped within amber is more generally a source of information about life in the so-called amber forest. Many coniferous remains have been found in Baltic amber, suggesting they had a place in the landscape, but this may also result from the fact that small and light needles are easily trapped. Fairly recently, scientists have linked some of the coniferous specimens to modern species of cypress and pine now growing in North America, China, Japan and Africa. Flowering plant inclusions are especially diverse but proportionally less common. Specimens of willow, date palm, sorrel, hydrangea, geranium, juniper, magnolia and camphor have also been found. These can be related to organisms growing in subtropical and tropical climates in southern Europe and North Africa today. Scientists today think that the climate in what is now the Baltic must once have resembled places where tropical, subtropical and warm temperate plant species co-exist, like present-day southern Florida or northern Myanmar. Indeed, some of the closest relatives to the insects found in Baltic amber today live in Southeast Asia, southern Africa and South America.

Scientists have also recently become aware that fossil resins preserve details of resin chemistry very well. Organic geochemistry can suggest the parent plants of some ambers by establishing relationships

11 Kauri (*Agathis australis*) in Dunedin Botanic Garden, New Zealand.

between fossil and modern resins.[28] The sources of some ambers have been clearly identified. Amber from the Dominican Republic, for example, is believed to have been extruded by an extinct taxon of tree known as *Hymenaea protera*, based on both chemistry and associated plant remains. Amber from Myanmar and New Zealand, on the other hand, has been linked to the genus *Agathis*, with *Agathis australis*, also known as the kauri 'pine', being the best-known living descendant today (illus. 11). The chemical approach has yet to bear conclusive proof of the source of Baltic succinite. Some scientists have suggested a relative of the *Pseudolarix* genus, for these trees also produce succinic acid. Remains of *Pseudolarix* leaves and cones have been found in amber from the Canadian Arctic Circle, confirming that they were growing (and producing resin) at high latitudes around the same time that Baltic succinite was being formed. *Pseudolarix* only produces resin when injured, supporting the idea that the huge quantities of amber found at the Baltic may have formed in response to external factors.

Chemistry may also help explain why some types of amber preserve inclusions better than others. It has long been observed that inclusions in Baltic amber have a milky layer, called *Schimmel*,

dangers involved but cannot dissuade the youth. As Phaethon has neither the experience nor strength to control its steeds, the chariot careers too close to the earth. Jupiter averts further destruction with a thunderbolt, knocking Phaethon, chariot and horses out of the sky. Phaethon plummets to his death on a riverbank and is buried by the river god Eridanus. When his mother and sisters finally reach his resting place, their grief is overwhelming. It transforms them. Their arms, thrown up in anguish, stiffen and are enveloped by bark; their fingers sprout leaves; and their tears harden to amber. These tears drop into the water, 'borne onward' by the river to 'adorn the brides of Rome'.[3] Significantly, death and transformation are also themes found in some Chinese excursuses. The Ming dynasty (1368–1644) writer Li Shizhen, for example, writing in the sixteenth century, explained to his readers that amber was the souls of dead tigers that had penetrated the earth and petrified.[4] This association with the fierce tiger means that amber is also associated with courage in Chinese culture.

PHAETHON'S RENAISSANCE RESURRECTION

The story of Phaethon was well known to the ancient Romans. Frequenters of chariot races, they were familiar with horses and chariots in full career, not to mention track-side collapses. The story was told and retold down the centuries but was especially popular in the period of European history referred to as the Renaissance (very broadly 1400–1600). In the hands of Italian humanists, Phaethon's death took on a moral dimension. Rulers, those who held the reins of government, were to heed his tale. Lodovico Ariosto employed Phaethon in an allusion to the 'political and moral problem of over-reaching one's station'.[5] For others, 'his case admonishes those who do not exercise control over their own impulses'; Phaethon 'died, not without example to all men'.[6] The dramatic tale was also perfectly suited to the art of the period. It was a natural candidate for the decoration of ceilings (illus. 14), a fact first noted by the second-/third-century writer Philostratus, who had seen Phaethon's fate depicted in an allegorical representation.[7] The episode was also used

14 Giulio Romano, *The Fall of Phaethon*, 1526, ceiling fresco,
Chamber of the Eagles, Palazzo Te, Mantua, Italy.

to decorate tapestries, furniture and ceramics. A surviving design for a
silver basin to be used with a ewer for handwashing at table highlights
that the tale could be paired with water as easily as with air.

ERIDANUS – PADUS – PO

Already in Pliny's time a body of writers had shown themselves keen
to locate the actual site of Phaethon's demise. There was little consensus. Some argued for Iberia, linking the name Eridanus (sometimes
spelled Eridanos) with the river Rhodanus. Others argued for the
Padus, the modern river Po. The Po reached the Adriatic at the point
where some authors claimed there was a cluster of islands known as
the Electrides, a name echoing the Greek ἤλεκτρον (*elektron*, the ancient Greek word for amber).[8] There were even advocates of Ethiopia.

15 François Duquesnoy or follower of, *Bacchante at Rest*, dated 1625, amber.

Pliny dismissed them all on the basis of 'such monstrous ignorance of geography', but these debates were slow to die in Renaissance Italy, where the Po was key to commerce, agriculture and defence.

In his *Descrittione di tutta Italia* (Description of All Italy, 1550), the historian Leandro Alberti assembled evidence to the effect that the Po and the Eridanus were the same. The Po ran close to the Este-ruled city of Ferrara and scholars there drew upon the Phaethon/Eridanus motif to establish an irrefutable link between contemporary court culture and the world of the ancient gods. In 1545, the Este dynasty, for whom the Phaethon story played an important part in their own mythology, had the relationship visualized in a series of five tapestries the scenes of which were inspired by Ovid's work and one of which depicted the transformation of the sisters into trees.[9] The Este were also among the first Italian nobles to develop extensive collections of amber. Their relationship with the Po and its amber was even indirectly responsible for one of the most curious amber artworks ever produced: a sleeping female Bacchante (a female follower of Bacchus) attributed to François Duquesnoy, a Flemish sculptor working in Rome (illus. 15).[10] The carving closely resembles a reclining figure in Titian's *The Andrians* (1523–5), a painting originally commissioned by Alfonso d'Este in celebration of the river (illus. 16).

Titian's *Andrians* has many curious details, including an unexpected guinea fowl. These birds, in fact, also have a place in amber myth. According to Greek mythology, the baby Meleager was predicted to live only as long as a piece of wood in the family fireplace remained unburned. To prevent this, his mother Althaea removed it from the fire. Years later, following the news that Meleager had killed

16 Titian, *The Andrians*, 1523–6, oil on canvas.
The guinea fowl can be seen in the tree right of centre.

17 Santi di Tito, *The Creation of Amber*, 1570–71, oil on slate.

her brother and another of her sons, Althaea returned the piece to the flames. Like the Heliades, Meleager's sisters, grieving his death, were transformed, this time into guinea fowl, and Sophocles wrote that amber, which he believed came from beyond India, was their tears. Pliny doubted Sophocles. 'Birds that weep every year or shed such large tears, or that once migrated from Greece, where Meleager had died, to the Indies to mourn for him' was too fantastic for him; others paid the idea some attention. The eminent Greek geographer Strabo, for example, brought the story closer to home. He maintained that the birds' habitat was the 'Electrides Islands that lie off the Padus'.[11]

DROPS OF DEVOTION

In the West, female devotion and the motif of tears is a common feature of many accounts of amber. Today the motif of women weeping tears of amber is a well-established literary trope. Victorian poets in particular embraced it. It gave rise to such titles as Thomas Holley Chivers's *Memoralia; or, Phials of Amber Full of the Tears of Love. A Gift for the Beautiful* (1853) as well as Elizabeth Barrett Browning's poem 'Comfort', in which she imagines herself as the Virgin Mary, praying that her tears will fall as amber at the foot of the True Cross of Jesus Christ.

The idea of sisterly grief has also compelled the creation of exceptional artworks. The most well known is a panel painted by Santi di Tito for the *studiolo* of Francesco de' Medici, Grand Duke of Tuscany, around 1570–71. It is unusual in that it goes beyond showing the transformation of Phaethon's sisters into trees and their tears of amber, for it also depicts the act of gathering the tears from the water at their feet (illus. 17). One of the women depicted has been identified as Lucrezia de' Medici, Francesco's sister, who had died shortly after marrying in 1558. This picture is therefore the complex expression of a brother mourning a sister through the depiction of sisters mourning a brother.[12]

A second well-known narrative linking amber's formation to female experience is from Lithuania. It became popular in the nineteenth century when Europe was in the throes of romantic nationalism. Liudvikas Adomas Jucevičius, a collector of folktales, recounted the story in his *Jūratė and Kastytis* (1842). The prince of the seas, Kastytis, abducted the mortal Jūratė, married her and crowned her with a coronet of amber. Jūratė consoled her grieving parents by casting nuggets of amber up on the tide. Another version tells of Kastytis the fisherman, who has created havoc by casting his nets where he should not. Jūratė, goddess of the sea, sets out to confront him but falls in love and takes him back to her amber palace beneath the waves. This enrages Perkūnas, chief of the gods. Perkūnas kills Kastytis, destroys the palace and chains Jūratė to its ruins. In yet another version, Jūratė is so enamoured with Kastytis that she leaves the sea to be with him. Perkūnas kills Kastytis, destroys Jūratė's palace and chains her to its ruins.

Like the story of Phaethon and the Heliades, which in the modern age inspired technology in the form of the phaeton open-top carriage, the phaeton automobile and the Volkswagen Phaeton, the tale of Jūratė and Kastytis has had a rich and romantic recent life. Its popularity began to grow after the unification of the Klaipėda (Memel) region with Lithuania in 1923, which brought a greater number of Lithuanians into closer contact with the amber coast, its people and its traditions.[13] Around this time, the patriotic poet Maironis reimagined the tale as a ballad. It went on to be staged as a ballet in 1933, and the artist Vaclovas Ratas produced an acclaimed series of woodcuts illustrating it in 1937. In the 1950s, the story was adapted to be sung as an opera (1955) and made into an animated film, Gražina Brašiškytė's *The Amber Castle* (1959). In the 1990s it became a play, and in 2002, marking the 750th anniversary of the founding of Klaipėda, a rock opera. More uncomfortably, however, the myth has been mobilized by the far right, for example in the lyrics of the metal band Diktatūra.

NO AMBER AT THE ERIDANUS

In northern Europe, the Phaethon story remained just that. It was treated as a myth by Prussian and German scholars. Their lack of interest probably resulted from the fact that the tale could not be related to the geography and topography they knew. Martin Zeiller was an exception. He tried to test the tale against reality, and he proposed that the name Eridanus might derive from the river Radaune (modern Radunia) in Prussia. He also thought it possible that the Vistula had once been called the Eridanus, and he suggested that land between it and the river Nogat was the Electrides. He proposed that the Samland Peninsula was the Glessaria of which the ancient authors had written (illus. 18).[14]

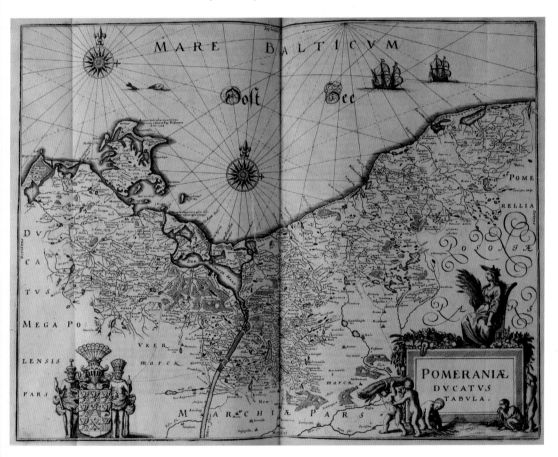

18 Map of Pomerania, from Martin Zeiller, *Topographia electorat, Brandenburgici et ducatus Pomeraniæ* (1652). This map focuses on Stettin (modern Szczecin) and the Oder. Danzig (modern Gdańsk) and the Radunia are further east. The Samland Peninsula can be seen at the easternmost extreme of the coastline.

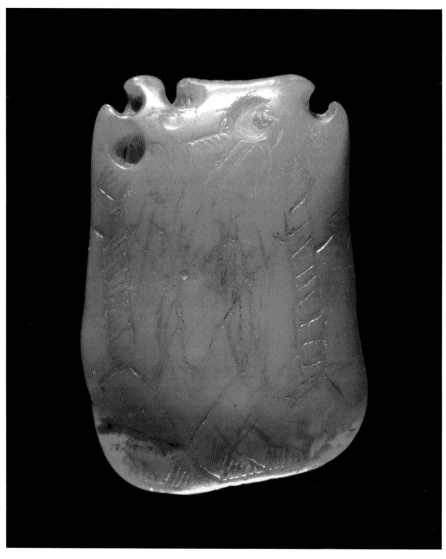

21 Pendant incised with human figures, from Sindalgård,
Åmosen, Denmark, 12800–1700 BCE, amber.

three

Ancestors and Amber

The oldest known discoveries of amber associated with human activity date back about 40,000 years, to a time when people lived by hunting and foraging using stone, bone and antler to make their sophisticated tools and weapons. From this time forwards, hunters, farmers, nomads, town dwellers, potters and metalworkers were attracted to and valued amber. The fascination with it probably inspired an array of meanings which varied according to use. How did people engage socially, emotionally and spiritually with amber? Archaeology gives us a range of information about innovations, traditions and historical awareness.

ICE AGE AMBERS IN EUROPE

In what is now Europe, the people of the late Ice Age, between 40,000 and 12,000 years ago, lived in tundra-like conditions that were colder than today but not always snowy or icy as the name of this period suggests. Though some were settled, it was mainly mobile hunter-gatherers who, for the first time, used amber (illus. 21). At the famous cave of Isturitz in the western Pyrenees, fragments suggest that amber was known to the people there around 34,000 to 29,000 years ago. Analysis has ruled out a Baltic origin, and it has been suggested that the amber may have been sourced at nearby Dax, where lignite used to make other personal ornaments found in the cave was also discovered.[1] Similarly, the amber unearthed in the caves at Aurensan in the Upper Pyrenees, inhabited 17,000–11,500 years ago, is likely to have been local. Recent research in Spain has also shown that local material was topped up with imported amber from southern Italy.[2] Around 11,000 years ago, early inhabitants of North America used amber-based adhesives to fix stone spear points to shafts and more recently the Thule ancestors of the Inuit turned locally sourced amber into beads.

In the region that has become northern Europe, Baltic amber deposits began to emerge from beneath the great sheets of ice as temperatures rose around 12,500 years ago. Until about 8,500 years ago, Britain was a western peninsula of continental Europe. As sea levels rose, the English Channel formed and water flooded the area now known as the Baltic Sea. Pebbles of amber washed up along the newly formed English coastline may have been the source of the Baltic amber excavated at Cheddar Gorge, Creswell Crags and Star Carr.[3] Today Baltic succinite is carried on the currents and washes up on Britain's eastern seaboard, but Britain also has its own natural amber, including amber found buried deep in London clays, which is still being excavated today by subterranean infrastructure projects.[4]

THE EARLIEST AMBER ART

Hunter-gatherers of the late glacial to early post-glacial period (40,000–10,000 years ago) in present-day Denmark also chose to pierce and decorate pieces of amber with dimples and line patterns, making beads and pendants.[5] One piece from Åmosen, West Zealand, is incised with human figures, making it one of northern Europe's oldest known depictions of people. Large nuggets carved into birds, elk, bears, boars and horses have been found, many washed up on the coast, churned up by the waves from now-submerged sites.[6] In 2015, one lucky winter beachcomber found a carved elk among seaweed on the beach at Næsby (illus. 22). The chance discovery caused a press sensation and it has been heralded as Scandinavia's oldest 'work of art'.

These little animal sculptures found as stray discoveries can only be dated by deduction, but the discovery of pieces of an elk figure

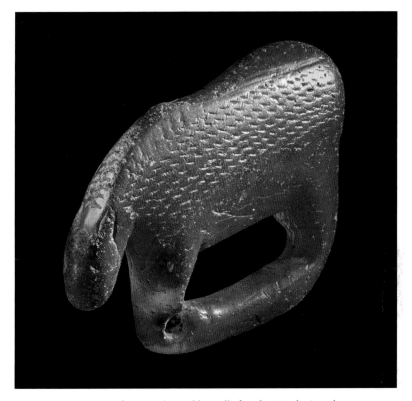

22 Figure of an animal, possibly an elk, found on Næsby Strand,
Denmark, 12800–1700 BCE, amber.

at a site excavated at Weitsche in Germany between 1994 and 2004
supplied other material that gave an estimated age of about 13,700
years old. The find was used to redate other figures, and these are
now stated to be the oldest known animal representations ever found
on the North German Plain.[7]

What is the picture like elsewhere? In Japan, amber from Chōshi
has been exploited since the Middle Jomon period (5,500–4,500 years
ago).[8] The earliest hunters and foragers were particularly interested
in the material as one of prestige. They transformed amber into
beads and pendants both to wear and to give as gifts that supported
the development of personal and social relationships, sometimes
over long distances. Such tokens of friendship are thought to have
facilitated access to hunting rights in other territories. The forms

these ornaments took may also have been related to magical beliefs as well as to the attempts of people to create and shape identities for themselves.

Wherever they occur in the world, most finds of ambers collected and modified by humans are made by chance or during archaeological excavation. However, one of the most famous finds ever made resulted from attempts to mine amber commercially in the later nineteenth century.[9] At Schwarzort, now Juodkrantė in modern Lithuania, the shallows of the lagoon were mechanically excavated. Sand and sediment were filtered through sieves constructed to trap even the smallest fragments of amber. The result was a haul of mysterious

23 Human figure found at Schwarzort, 2350–2050 BCE, amber.
Now in the Geowissenschaftliches Museum, Göttingen, Germany.

anthropomorphic figures and forms, which are now thought to have been made about 4,350–4,050 years ago, roughly around the era when Egypt's pyramids were built (illus. 23). These artefacts were initially treated as curios and given away to employees and visitors; later they were assembled in the on-site museum. Only a fraction of the original objects survived the Second World War, having been dramatically evacuated to Göttingen, Germany, when Prussia was occupied by Soviet forces.

The exact purpose of such figures is still unclear. Amber animal carvings found in Denmark and elsewhere are also worked with markings. Are these intended to bring greater life to the carving or are they auspicious – or could both be true? The fact that the figures have sometimes been effaced and reincised suggests that these representations may have had some form of agency and functioned as hunters' amulets. The figures found at Schwarzort are often decorated with scatterings of drilled dots, which suggest detail and texture. With their shield-shaped faces, wide-bridged noses and simply rendered mouths and members, they have been described as owl-like, and modern scholars have suggested they represent animal spirits.

With their deep orange tones and smooth, lustrous surfaces, these objects use amber in arresting ways. They represent the application of skill and imagination to create items of beauty and emotional power. Their making required larger pieces of amber and they certainly had an inherent material value which is difficult to quantify today. Where the material had travelled far, this value must have been all the greater. One of the oldest known finds of Baltic amber beads in the Near East was made in Assur on the Tigris river in northern Iraq. It dates to circa 3,800 years ago.[10] A spectacular lion-shaped amber vessel dating to about 3,360 years ago was unearthed in the excavation of a royal tomb on the site of the ancient city of Qatna, Syria (illus. 24). When tested, this amber was revealed to be succinite (Baltic amber) and the lion is the oldest known figurative treatment of amber outside of Europe. Archaeologists suggest that it was worked locally and is likely to have been imported in its natural state, probably over the Aegean Sea, a significant trade route.[11] Most other ambers, usually beads,

as many as five hundred beads. One of the most remarkable finds in recent years has been of a burial at Bettelbühl in Germany, thought to be of a princess or priestess aged between thirty and forty and possibly her daughter, buried around 580 BCE. The grave chamber was filled with highly crafted jewellery, including some one hundred items in amber (illus. 27).[24] This grave is the earliest to have been found of a woman from the Celtic elite. About a hundred years later, not so far away, an equally sumptuous burial took place, and this amber shows quite different influences. The man laid to rest around 500 BCE at Grafenbühl, Germany, was interred on a splendid bier decorated with inlaid amber palmettes thought to have been made in southern Italy. His grave also held two amber-faced Sphinxes made even further away in Tarento, a Spartan colony in the Greek Peloponnese.

THE AMBER ISLAND

The Celts are possibly the people living 'beyond the North Wind' who Herodotus described as being responsible for the amber at Delos. Herodotus also referred to a river 'flowing into the sea which lies towards the North Wind'. He had heard that amber originated there.[25] His is one of many texts which converge on amber as the product of a mysterious northern ocean, one of the earliest being a cuneiform tablet from circa 4,000 years ago explaining 'in the sea of the changing winds, merchants fish for pearl; in the sea where the north star culminates, they fish for amber'.[26]

Had Herodotus lived a century later, he would have been able to benefit from the experience of Pytheas, who hoisted his sail in what is today Marseilles and set off for northwestern Europe. According to the much later Pliny the Elder, Pytheas told of an island called Abalus, 'upon the shores of which amber is thrown up by the waves in spring'. The region's inhabitants, people Pytheas called the Gutones, used it for fuel and traded it with the Teutones, their neighbours.[27]

Scholars have debated which of the Baltic's many islands is Pytheas' Abalus, for to identify it would be to know the mariner's route. Some have argued that the name Abalus derives from the

26 Carved fragment of an Etruscan fibula (dress pin),
known as the Morgan Amber, *c.* 500 BCE, amber.

27 Amber jewellery found during the excavation at Bettelbühl of the grave of an unidentified
Celtic princess, near Herbertingen, Baden-Württemberg, Germany, *c.* 600 BCE.

Celtic word for apple and point to Celtic legends of an island forested with trees whose fruits fall as amber. Others have highlighted that the Gutones were the inhabitants of the Swedish island Gotland. A contemporary explorer called Timaeus gave an alternative name for Abalus. He said it was also known as Basilea. Today, there is some consensus that Basilea is Helgoland. Lying west of Jutland and at the head of the river Elbe, Helgoland was the mythical realm of the Frisian god of the dead and its name means Holy Land. Basilea is probably its Hellenization.

TRAVELLING AMBER

Jutland is likely to have been one of many regions encircling the Baltic Sea from which Baltic amber was sent onwards. Trade-route mapping has long been a core activity of amber studies (illus. 28). It is now thought that amber was shipped southwards inland on the rivers Oder and Vistula before following different land and river routes to reach Hungary and the Black Sea. It could have descended the river Elbe to its confluence with the Vltava and crossed southern Bohemia to Linz and the river Danube. From here it would have been able to reach markets in Austria, Hungary, Germany and Switzerland, and would only have needed to cross the Alps to arrive in Italy and at the Adriatic coast.[28] These routes continue to fire the modern imagination. They have inspired caravanning guides, postage stamps and even coins. In Poland, the country's major north–south highway is known as the Amber Highway. In Austria, where Roman Carnuntum was a staging post for amber on the way to Aquileia, children can follow an amber-themed treasure hunt, eat at restaurants offering an amber-related menu and sing along to an amber musical (illus. 29).[29]

More research is still needed if the mechanisms and transactions that conveyed amber throughout Europe are to be explained. Artefacts found in Bohemia, Austria, Romania, Albania, Serbia, Greece, Slovenia, Croatia and Italy are often similarly shaped, but why they are so similar is yet to be fully understood. With time, it may become possible to ascertain whether the artefacts reached more

28 Ancient and medieval routes for the trade of
amber from the coast of the Baltic Sea onwards.

29 Betty Bernstein 'the girl with the magic piece of amber', figurehead of the 'Betty Bernstein' tourism initiative, www.betty-bernstein.at.

southerly consumers directly, via intermediaries or through personal migration. Questions are also being asked about routes northwards to Sweden and Finland, as well as further westwards to Britain and Ireland. How is it, for example, that characteristically English types of ambers have been found in southern Germany and the Minoan Mediterranean? To date, European scholars have had little to say about how amber reached Asia. This is because there is little printed in English or other European languages about amber in places such as China during the same period.

AMBER IN ANCIENT CHINA

The oldest amber art in China has been found at the Sanxingdui site in Sichuan. Sanxingdui culture flourished around 3,200 years ago (c. 1300–1200 BCE), making its people rough contemporaries of the already-discussed European Mycenaeans. The oldest object known is a heart-shaped amber pendant carved with a cicada and branches that was excavated from a sacrificial pit.[30] Finds dating to the subsequent Shang (1600–1046 BCE), Spring and Autumn (771–480 BCE) and Warring States (480–221 BCE) periods are few and rare. Most are single beads, but there are also exceptional carved pieces, such as the amber tiger buried with an urn containing a child's remains in Tangshan.[31]

In China, the history of the consumption and appreciation of amber first begins seriously during the Han dynasty (206 BCE–220 CE). Artefacts and anecdotes survive which underline the material's great prestige and show that it was almost exclusively enjoyed by the royal family and high-ranking nobles. Some contemporary authors briefly address the forms into which it was worked. Amber pillows stand out, though none survive.[32] Instead, small ornaments shaped into tigers, frogs, turtles, ducks, butterflies and rabbits have all been found, as have personal seals copying forms in bronze.[33] These animals are linked to luck and protection and traces of wear suggest they had a useful rather than completely symbolic life. The contexts of their discoveries and the fact that many are perforated suggest that many may have been worn, some in series with jade and pearls as necklaces or in the hair as headdresses.[34] Fascinatingly, the discovery of an amber bead showcasing inclusions in the celebrated tomb of the Marquis of Haihun, sealed since his death in 59 BCE and excavated in 2011, demonstrates that the Western Han also treasured amber's ability to preserve insect life.[35] Han ambers have also been found west of China, for example at Tillya Tepe in Afghanistan.[36]

AMBER AND ANCIENT ROME

Most European histories of amber begin with the Roman era. The first extensive written accounts of amber in Latin date to this epoch. They not only helpfully summarize now-lost written works of previous centuries but cover the contemporary fascination with the material in depth. The most well known was written by Pliny the Elder, a statesman, military commander, philosopher and natural historian, who died in the eruption of Vesuvius that destroyed Pompeii two millennia ago in August 79 CE.[37]

Pliny dedicated two full chapters to amber. In the first he coun-tered 'falsehoods' concerning where it comes from and what it is. In the second he dealt with types of amber and amber's use. It was not a material which he personally wanted to own. He considered objects made of amber a wasteful luxury. He had seen it used to stud the nets employed to contain animals in amphitheatres and to decorate weapons, seating and other equipment. He felt it had no practical or sensible application and, worst of all, it was particularly coveted by women. He was disgusted that little amber carvings commanded more money than enslaved people. Juvenal, writing a little later, mocked those who spent money on amber, laughing at one millionaire who allegedly ordered 'a troop of slaves to be on watch all night with fire buckets' so anxious was he 'for his amber, his statues and Phrygian marbles, his ivory and plaques of tortoise-shell'.[38]

Pliny had heard about fellow Romans who had been to the Baltic. He was informed that the people of the region referred to amber as *glaes* (Latin *glaesum*), related to the modern word 'glass' today. They also talked about an island: Austeravia (*rav* means amber in Danish and Norwegian, from the Norse *raf*, meaning fox), known also as Glaesaria. About twenty years after Pliny's death, Tacitus wrote his account of the region. He noted that the Aesti, living on the 'right shore of the Suebic sea' collected *glaes* in the shallows and on the beach:

> As usual with the barbarians, they have neither asked nor
> ascertained its nature or the principle that produces it; quite
> the contrary, it lay long unnoticed amidst the other jetsam of
> the sea, until our extravagance gave it a name. To them it is
> utterly useless: they collect it crude, pass it on unworked,
> and gape at the price they are paid.[39]

The Romans constructed themselves as culturally and technolog-ically advanced vis-à-vis the peoples of Germany, instrumentalizing amber in this. Roman coins found in cemeteries in Samland demon-strate the extent and length of Roman contact with the region well into the third century CE. A general lack of worked ambers found to date may suggest that the local people had different patterns of consumption. The ambers so prized by the Romans were not worked in the Baltic but in Aquileia, at the head of the Adriatic. This town was a major centre for working amber and from which intricate and beautiful carvings were sent out across the Empire.[40] Its amber craftspeople made handles for cutlery, perfume pots, playing pieces, dice for gaming and gambling, combs, miniaturized implements for weaving and spinning, pendants and figurines of the Roman gods, particularly of Cupid and the Lares (see, for example, illus. 20).

AMBER IN FIRST-MILLENNIUM EUROPE

When the Western Roman Empire faltered and collapsed, the political complexion of the continent changed dramatically.[41] Between about 1,700 and 1,200 years ago, Europe's different peoples moved from and settled beyond their historic heartlands – a period historians call the Great Migrations. Between about 1,800 and 1,500 years ago, consumption of amber by people local to the southern Baltic shore grew. Abroad it remained as associated with them as ever, not least thanks to their use of it in their diplomacy. The sixth-century statesman Cassiodorus recalled the arrival of an Aesti delegation in Ravenna with a gift of Baltic amber for Theodoric, king of the Ostrogoths. The gift seems to have been a disaster. The Aesti claimed to be unable to say more about the specific origin of the material, generating the impression that they were hiding something.[42]

The Ostrogothic Kingdom was just one of the powerful territories which developed in the vacuum left by Rome. In North Gaul (modern Benelux and German Rhineland) there were the Franks ruled by the Merovingian kings and, in England, the Anglo-Saxons. All utilized amber – primarily in bead form. The numbers are striking. A recent study of about two hundred Merovingian grave sites has highlighted 5,500 single incidences of amber, meaning nearly one in every ten beads found in Merovingian interments is amber, and nearly one in every three graves containing beads has amber.[43] In early medieval

England, amber is often combined with clear rock-crystal. In one case, modest amber beads have been strung either side of a heavy rock-crystal spindle whorl so great that the piece had to be worn suspended between two fibulae. Amber sword beads, single beads affixed to swords, show the use of amber in the protection of warriors, as well as in bodily ornament.[44] These demonstrate elements of continuity and universality in the relationships people share with amber.

AMBER IN FIRST-MILLENNIUM CHINA

In Three Kingdoms China (220–280 CE), continuity expresses itself in the type of people able to decorate themselves with amber. Members of the royal family and nobles continued to be buried with amber, just as they had been during the Han dynasty (206 BCE–220 CE), when amber first began to be significantly used.[45] Yet the first millennium was also a period of innovation in Chinese ambers. The extensive use of amber in inlays is not seen in Western Europe until the twentieth century – in China, the earliest surviving example of this technique, thought itself to have been inspired by Scythian examples, comes from the Eastern Wei dynasty and the tomb of a young Rouran princess (d. 550 CE).[46] Inlaying amber expanded under the Tang (618–907 CE), and amber is found inlaid into objects such as censers. It is also used to make receptacles and drinking utensils.

Though burmite (amber from Myanmar) was known to the later Han, it seems that much of the amber consumed in China in this and later eras was Baltic in origin. The Book of Wei (551–4) and the Books of Liang (635) and Sui (636), as well as the Book of Tang (c. 941), all name places in central and western Asia as sites of supply, and towns such as Dushanbe and Samarkand were situated on trade routes which carried Chinese porcelain and silk to Europe and brought European materials to China. Sources also refer to the importance of Daqin, the ancient Chinese name for the Roman Empire, or at least the part of it known to them, which is roughly modern-day Syria. Did Tang (618–907 CE) and Song (960–1279 CE) poets know of a similar tradition in Roman Europe when comparing the colour of amber to that of yellow wine from Lanling, a drink made by fermenting rice, barley or corn? Whatever the answer, the use of amber as a comparator suggests that the material was known well enough for the reference to make general sense.[47]

Amber was not only transformed into objects; it was valued for health-giving properties. It has been found in the silver capsules that are part of the late eighth-century Hejiacun hoard in Xi'an, alongside drugs, aromatics and other substances classed among the Seven Treasures of Buddhism. Amber is sometimes given as one of the religion's sacred stones and is also held by some to symbolize Buddha's blood. It has therefore long played a role in Buddhist devotion.

Some objects have fascinating tales to tell. Little amber cats were found in the crypt beneath the temple constructed to honour a relic of Buddha at Famen in Xi'an, sensationally discovered after rains washed away part of the structure. Comparison with examples of the same subject excavated in what was Roman Europe has led to the suggestion that these are of European facture and use Baltic amber. It is assumed that they were given to the Tang court as a gift, possibly via northeastern China or Mongolia. This region would soon become intensely important for the history of amber in China. When reading about amber in first-millennium China it is neither Tang nor Song ambers which receive the lion's share of attention, but rather ambers pertaining to the Liao dynasty (907–1125), whose great state took in Inner Mongolia and swathes of north and northeastern China. Contemporaries of the Vikings in Europe, the Liao, or Khitan as they are also known, were at the height of their military power around 1000.[48] They were also superbly placed, connecting the Song to their south, Korea and Japan to their east, the peoples of Siberia and the Kirgiz to their north, and the Abbasids, Samanids and Qarakhanids, among others, to their west.

Amber is found in considerable quantities in Liao-period tombs (illus. 30, 31). A staggering 2,000 pieces were found when the tomb of the Khitan princess Chen (d. 1018) was excavated in Qinglongshan. The princess and her consort were buried with amber jewellery, amber perfume containers and amber-handled hunting equipment.

30, 31 Liao dynasty (907–1125) ornamental plaques, 11th–12th century, amber.

Testing has revealed much of this amber to be Baltic in origin, and this is thought to have been obtained not only via what are now anachronistically known as the Silk Roads but via the more northerly Fur Routes plied by the Kyivan Rus'. The Rus' trafficked amber from the Baltic via Novgorod and sent it southwards along the rivers Volga and Dnieper into the Byzantine Empire and Arab world. They also traded it across land to the Caucasus and the regions' khanates. Though amber is now known to occur in small quantities in Lebanon, for centuries the only amber to be easily had in the Near and Middle East was Baltic.[49]

AMBER IN CAUCASIA AND WESTERN ASIA

The Vikings, who worked Baltic amber at Ribe, Hedeby and Birka, were early lynchpins in amber's journey eastwards, exchanging it for Russian furs in places such as Ladoga. From there merchants sent it onwards to Iran and Iraq. By the eleventh century, amber was sufficiently well known among elites to be referred to in verse written at the court of Mahmud of Ghazni. The poets Unsuri and Farrukhi Sistani compare tears to amber and celebrate blood spilled in military campaigns as being like amber beads.[50] Later, the thirteenth-century Persian poet Rumi would refer to its power to attract little bits of straw when rubbed as being like the attraction of lovers for one another.[51] This curious feature of amber is enshrined in the Arabic *kahroba* and Persian *kahraba*, which mean something like 'straw robber' and refer to its ability to attract light matter.

The source of this amber was poorly understood. A rough contemporary of Rumi wrote that the amber available in Khorāsān, northern Iran, had come from the sea via eastern or central Europe.[52] There is growing evidence for the widespread availability of Baltic amber in western Asia in the latter centuries of the first millennium and first centuries of the next. Recently archaeologists excavating properties destroyed in the Mongol invasions of the Rus' territories have added to knowledge and revealed something of its extent. In Vladimir, around 195 kilometres (120 mi.) east of Moscow, some

Faith is more explicitly expressed in such objects as chalices: the vessel Christians use to celebrate the Eucharist, a rite in which they celebrate Christ's martyrdom by drinking wine symbolic of his blood. Amber has been associated with wine since Roman times and its colour compared to the famous white wine produced on the slopes of Mount Falernus in Italy, between Rome and Naples, which turned reddish brown after ageing in amphorae for twenty years. In Ireland, amber was used to stud the Derrynaflan chalice and to link literal wine with a material the colour of which was wine-like. Amber was also used to embellish the bindings of Gospel-books, which were themselves the physical embodiment of God's word. These belonged with chalices and other liturgical apparatus in church treasuries, and exquisite bindings emphasized that manuscripts were there not only to be read, but to be seen.

TREASURING AMBER

Almost nothing is known about the artisans who made these objects and where, when and why they sourced amber pieces. But there is information about the objects themselves. The Derrynaflan chalice was buried for safekeeping during the turbulent tenth to twelfth centuries. The Hunterston brooch is inscribed with a woman's name, presumably its owner. She was called Melbrigda, a common Irish Gaelic name associated with the cult of St Bridget. Yet interestingly the letters and words are written in Viking runes and Old Norse, and even more tantalizingly the brooch was found buried in Scotland. The Viking relationship with the Rus' was key to the arrival of Baltic amber in East Asia and the Norse themselves were great amber-lovers. Amber workshops existed in Viking Dublin and York, supplying both the new Scandinavian and established local populations with beads, pendants, many shaped like hammers and axes probably related to Thor, figurines and games pieces.[56] Some have a captivating power. One little gaming piece in the form of a man sagely stroking his long beard has been calmly contemplating for centuries.

32 Brooch found at Hunterston, Ayrshire, Scotland, in the 1830s,
c. 700 CE, gold, silver and amber, made in Ireland or Western Scotland.

Humans have used amber from the earliest times. In its worked form, people have encountered it in many incarnations – some lavish, some unobtrusive. Certain objects were made to be treasured forever and passed from generation to generation, others were dedicated to the gods or spiritual leaders such as Buddha and Jesus Christ and taken out of circulation, and some followed their owners to the grave. Archaeologists and historians will probably never be certain of the individual motivations of makers and owners or of the ways they engaged with their ambers. They will also never know, in the absence of more written evidence, how those objects acted on or influenced them. This chapter has focused on what surviving ambers say about ancient and early medieval cultures; in later chapters, which benefit from greater surviving textual, visual and physical evidence, this balance changes.

four

Unearthing Amber

In Western literature, the Roman author Pliny the Elder was the first to discuss where and how amber was found. He focused on the far-away Baltic coast, where he had been told amber occurred (illus. 33). Pliny had been informed that its incidence differed with seasonal weather patterns, and that it was thrown up by the waves in the greatest quantities in spring.[1] Tacitus, writing slightly later, noted that locals actively foraged for it, scooping it up from shallows and picking it up from the beach.[2] Two millennia later the picture was little different. In 1929, Otto Paneth, a newly appointed professor at the university in Königsberg (now Kaliningrad), a town at the heart of the Baltic amber-finding region, wrote to his brother explaining how easily one happened upon it:

> One can find it oneself, without trouble, in innumerable little
> pieces, if one walks along the beach on certain days after a
> storm and is lucky enough somewhere to have caught the
> 'amber vein' which often continues for several kilometres at
> the same distance from the water, a line of seaweed deposited
> by the storm with small points of amber sparkling golden
> in the sun.[3]

Paneth noted that large pieces were rarer. For centuries, if these were chanced upon, they had to be ceded to the State by law. It had been this way since the Teutonic Knights arrived in the region. This

Two sources give information about amber collection in this period and both present some of the earliest evidence of active 'fishing' for amber rather than opportunistic foraging (illus. 35). The first is a discourse purportedly written by a local Dominican monk called Simon Grunau from the village of Tolkemit (now Tolkmicko) on the Vistula Lagoon, possibly around 1520:

> At night-time you see [amber] shining and floating around in the water; however, the biggest pieces are on the seabed. But if a storm blows in from the north, all farmers living nearby must come to the beach, and run into the sea with nets, and fish out the amber that floats up. As many bushels as one can fish, the same amount of salt is given to he who has collected them. Many farmers drown fishing like this.[9]

The second source is Georg Bauer/Agricola, a scholar based in one of the important ore-mining towns of Saxony, lying more than 600 kilometres (373 mi.) to the southwest. His field of expertise was the incidence, extraction and properties of minerals and metals. His wider work on these subjects, published in 1546, contains details about deposits of amber, as they were then understood, and exactly where amber was fished for. His focus is Samland and he writes:

> There are about 30 villages of the Sudini [his name for the people living there] who live on that portion of the promontory near Brusta [Brusterort, now Mayak] and these people today ... gather amber in small nets in the same manner as they catch fish ... practical experience has taught these people the best method for collecting this mineral and this knowledge has been passed down from hand to hand as they say.

When the north and northwest winds blow,

> The people ... rush eagerly from their villages, at night as well as during the day, to the beach upon which the waves

are driven by the winds. The men bring with them their nets woven from linen cord and fastened to the ends of long poles with prongs. When spread out these nets are as long as a man's arm. The women act as helpers. When the wind dies down, but with the sea still running high, the men, completely naked, run into the sea in the wake of each wave, and gather in their nets the amber which has been carried along the bottom. At the same time, they pull up the plants . . . They collect the amber and any plants as quickly as possible and when the next wave comes in they run for the beach where the wives empty the net . . . During the winter months of the year the wife warms the cold body of her nude husband with cloths that have been warmed by the fire so that he . . . can go back into the sea refreshed. He goes back again and again until he can find no more . . . He must take everything that he finds to an overseer who gives him an equal measure of salt for the amber.[10]

Both Grunau and Bauer record that amber was exchanged for salt. Since many of the farmers living by the coast were also fishermen, salt could be used to preserve their catch. However, the very fact that fishermen made their living from the sea also meant that they sometimes stood accused of only pretending to net fish but catching amber instead. Notices warned against amber's unlawful possession (illus. 36), and fishermen and farmers were obliged to swear that they would denounce even the closest family members. In later centuries, not only fishermen and farmers but their sons and their servants (if of age), postmen and priests were obliged to swear this oath and to repeat it every three years. They were also obliged to take part in campaigns to collect amber. At all other times, all Prussian beaches were out of bounds. Punishments ranged from fines, flogging and expulsion to death by hanging and they were meted out for the 'theft' of as little as 50 grams (1¾ oz).

Georg Bauer never visited Prussia and must have obtained his information about amber fishing from someone closer to the action.

36 Edict concerning the theft and smuggling of amber, dated 24 March 1664.

It was probably Andreas Goldschmidt/Aurifaber, the duke of Prussia's physician, who wrote his own even more detailed treatment of amber in 1551. According to Goldschmidt, refusing to collect amber met with a fine. He also explains that the nets used were a sizable 70 centimetres (27½ in.) wide. He is at pains to stress that naked fishing was not universally practised, but instead the custom of only a handful of villages spread across an area of about 5 kilometres (3 mi.)![11]

Though it seems an amusing aside, nudity may have been adopted to avoid drowning. In a woodcut from an eighteenth-century publication, netters are captured waist-high in the water (illus. 37). One nineteenth-century author said that amber fisherpeople in his time typically strode one hundred paces into the water or about as far as the third wave out.[12] Fishing in numbers was also safer. In the late 1600s, a single overseer controlled two beaches, each served by about 25 men. A single action might therefore involve as many as fifty collectors, with perhaps four hundred working in Samland on a single day. When the swell threatened to overwhelm the men, they turned sideways and banded together, creating a breaker of bodies. The net too was a lifesaver. They drove its handle into the sand and clung onto the pole, as the waves buffeted them on their way inland. It must have been a curious sight. In 2015, an English newspaper described the activities of the amber fishermen at Pionersky as a 'madcap treasure hunt'.[13]

There were safer methods too. Amber was also collected from boats. Long pikes were used to release amber from the seabed, and it was scooped up in nets or with tongs (illus. 38). Pincer-fished amber was worth twice the net-harvested variety because its condition was better. The safest of all was foraging for washed-up amber among the mouldering seaweed, and this also allowed for the involvement of women and children. This type of amber was commonly known as 'old amber' because it had been pummelled, shattered and buffeted

37 Amber fishing, woodcut from Johann Amos
Comenius, *Orbis sensualium pictus*, part II (1754).

by the sea. In 1537, payment for old amber was two-thirds less than for fished amber.[14]

Andreas Goldschmidt gives clear information about what happened after collecting had been completed. The then Amber Master was a man called Hans Fuchs. Fuchs arranged for amber to be packed into barrels and eventually delivered to Lochstädt (now Pawlowo). Once there, it was sorted into types. The three principal divisions were: common stone, turning and bastard. Goldschmidt does not

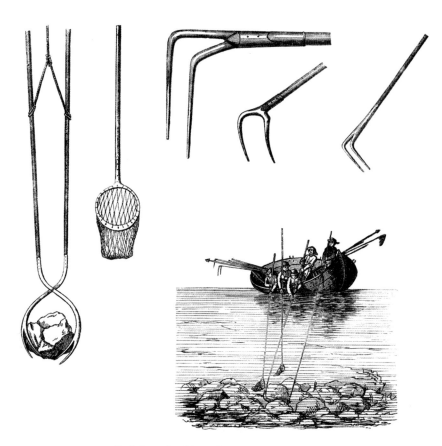

38 Tools for amber fishing, illustrated in Wilhelm Runge, *Der Bernstein in Ostpreussen* (1868).

relate what these terms meant in practice, but fifteen years later his compatriot Severin Göbel thankfully did:

> common stone is so-called because it is a mixture of the smallest and worst pieces; it is used by painters, carpenters, and others for varnish. The other is so-called turning stone from which you may turn or work all types of things. The third is called bastard; these are the largest pieces.[15]

A set of seventeenth-century instructions written during the tenure of the Amber Master Hans Niclaus Demmigern gives greater detail about size, colour and texture:

> Turning Stone: is that which is as long as a man's thumb and is as broad and thick as a man's thumb is long. It is brown or reddish. Ditto the brown or reddish pieces, if they are just over this size and are not good, firm but rather full of cracks, and holes, are bristly and 'worm-eaten'.

Bastard stone had the same dimensions, but:

> it is yellow, dark or light. If a piece of bastard is over a man's thumb long, broader than a man's thumb is long and thicker than a thumb but is 'worm-eaten', full of holes, bristly or unhealthy, it also belongs with bastard stone.

Demmigern's instructions also highlight a further category.

> Top Stone: is that which is broad, big and firm. Ditto, what is broader than a thumb's length and thicker than a thumb's length and also firm. [It is] the stone which is clear, light and firm and weighs 70 g or more, or, if it weighs less than this, is firm, fiery and clear. The stone which is white, cabbage or whey-coloured and also white bastard, whether it is big or small . . . belongs to top stone.[16]

Common stone was everything not covered in the above, apart from all white amber (which was reserved for medicines) as well as unusually coloured amber, which was to be singled out and taken to the Amber Master separately. Today the picture is remarkably similar. Amber is sorted into size according to weight and brought together for sale in weight-based categories. As of February 2020, miscellaneous pieces of amber weighing between 2 and 5 grams (0.07 and 0.18 oz) commanded around £450 per kilogram, and pieces of 500–1,000 grams (1–2¼ lb) around £4,300.[17]

Once sorted at Lochstädt, the barrels were transported to Königsberg and further onwards to Danzig, where a family of well-connected merchants by the name of Jaski organized its sale (illus. 39). The Jaski bought this right in 1533 for a first payment followed by set annual payments regardless of whether the agreed

39 Panorama of Danzig (modern Gdańsk), hand-coloured engraving from Georg Braun and Franz Hogenberg, *Civitates orbis terrarum*, vol. II (1575).

volume was supplied. Paul Jaski, the first in the line to acquire this right, was ennobled by the emperor in 1533 and had partners in the Low Countries. Several of his ten children went to university elsewhere in Europe and settled abroad. The Jaski went on to control the sale of amber for more than a century, despite many attempts to renegotiate made by the successors of the Teutonic Order, the dukes of Prussia. When, for example, in 1586, Duke Georg Friedrich sought to break their hold, the family appealed to their protector the king of Poland, who found in their favour. This hold was finally broken in the 1640s, when Elector Friedrich Wilhelm paid the monetary equivalent of 880 kilograms (1,940 lb) of silver to regain control. Payments were staggered over four years, and in 1647, Friedrich Wilhelm became the first ruler of Prussia to master amber from sea to sale.[18]

MAPPING AMBER

Several treatises written by local Prussians supply key sources of information about Baltic amber. But why did they bother to write about it? And why did collecting, sorting and selling amber become so important in the 1500s? The reason lies in the political and religious changes of the time. In 1525, Albrecht of Hohenzollern, Grand Master of the Teutonic Order, converted to Lutheranism. This compelled the creation of a secular state which he, and his heirs, would henceforth rule as dukes. These changes in status made Albrecht look at the territory, its boundaries, its history and its resources. Ducal Prussia, a territory now at the vanguard of the Lutheran reformation, was set on defining itself for others rather than being defined by them.

Albrecht surrounded himself with Protestant scholars. These were specially chosen for him by Philipp Melanchthon, friend of Martin Luther and a fellow reformer, from the faculty and students at the prestigious university of Wittenberg. Mapping the territory was a particular priority. The first maps highlighting amber and charting Samland and the Curonian and Vistula Lagoons date to Albrecht's reign. Olaus Magnus' huge 'Carta Marina' (1539) locates the amber coast through the inclusion of a solitary amber fisherman. Sebastian

Münster's map, now known to have been based on a map prepared to support a proposal for a geography of Prussia by the ducal astronomer Georg Rheticus,[19] was reproduced in every edition of his *Cosmographia* and tells readers: 'here you find amber.'

Understanding Prussian geography went hand in hand with understanding Prussian natural resources. Rheticus also investigated Prussian amber, which he discussed in terms of Aristotle's theory of stones. According to Aristotle, the sun caused the earth's humours to vaporize and then solidify. Rheticus highlighted that though the sun was low in northerly Prussia the generation of amber was not hindered and took this as evidence that the sun's rays were as effective when oblique as when beating down directly from above.[20] It was important that rulers understood the resources on their doorsteps, for the right to exploit them was exclusively theirs, and they could be keys to great wealth. In Rheticus' time, natural resources were not seen as chance occurrences but celebrated as extraordinary gifts of God. Similar words were recently used in an interview by a professor at Kaliningrad Technical University, who said: 'Amber is not a brand, it's a part of our life, very precious and special, that is given to our region as a gift.'[21] In the early 1500s, ore mining had made other German regions very wealthy and 'professionalization' of amber's administration and collection under Duke Albrecht relates to amber's potential to do the same. The development of a sophisticated system of grading is closely allied with a diversification of applications and greater marketability.

MINING AMBER

In sixteenth-century Europe there was no consensus that amber was a tree resin. Most Prussian writers thought that amber was actually some kind of pitch and that it was produced in the earth, hardening there or in the sea. The fact that it could be dug out of the dunes as well as fished from the sea was a crucial factor in these arguments. The dunes on the west coast of Samland are famous for a geological formation known as blue earth (*Blaue Erde*). The name refers to a layer containing glauconite and therefore of grey-green-blue appearance. This sedimentary stratum, which is believed to have built up in the delta of an ancient river, can yield up to 2 kilograms of amber per cubic metre (4½ lb per 35 cub. ft).[22]

Extracting amber from the dunes in the sixteenth and seventeenth centuries was an arduous process, for the 'stone [lay] one or two men deep in the earth' and miners were using little more than shovels.[23] Men on/in the dunes are shown in some maps of the period (illus. 40). They obtained licences to mine from the duke, paying in barrels of beer and part of their profits. The shafts and pits quickly

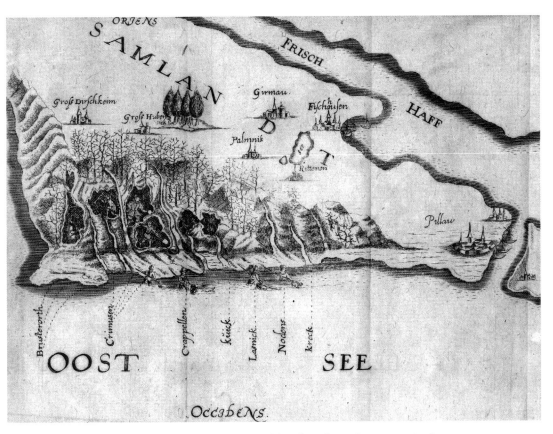

40 Mining for amber in the dunes and fishing for amber in the sea, engraving from Philipp Jacob Hartmann, *Succini Prussici physica & civilis historia . . .* (1677). Note the village of Gross Dirschkeim on the upper left. The dunes there are captured in the photograph reproduced as illus. 33. A view of Fischhausen, on the upper right, is also provided in illus. 9.

41 The Cuatro amber mine, located in the area
of El Valle, Dominican Republic, 2017.

filled with water, which was both good and bad, since according to
some accounts water washed amber from the sand. Mining in this
way does not appear to have been widespread, and attempts by one
landowner in 1649 to have his tenants dig for amber in the dunes
were rated 'highly unusual' by the Amber Master. The figures tell
a similar story: in 1670–71, only eight barrels were retrieved from
mining operations in Gross Hubnicken, Warnicken and Grünhof
(now Sinjawino, Lesnoje and Roszczino). In the years between 1705
and 1715, the same dunes produced just under nine barrels. By the
1760s, this method was considered thoroughly antiquated,[24] although
it still occurs today, particularly in the Masuria region of Poland and
where amber is mined illegally.

Makeshift pits, shafts and tunnels are by far the greatest yielders
of amber in other parts of the world too (illus. 41). A recent exposé in
an online journal described a contemporary mine in the Dominican
Republic in frightening detail:

When the scientists and I reached the amber mine it was
marked by a red tarp propped up to offer some shade over
its jagged mouth. It was little more than a hole in the ground,
its sides supported by bamboo poles at each corner. Inside,
140 feet below the surface, three men were hand-drilling
amber out of the surrounding rock. As we stood watching,
a young man beside us walked toward a stationary motorbike
whose engine was connected to a pulley. He revved the engine
and the bike sprang to life, puffing out a cloud of smoke. The
engine yanked the pulley, which raised a rope out of the mine
– pulling up first a bucket of fist-size chunks of amber, then
the workers themselves. One by one, they emerged, covered
in clay, barefoot, and shirtless.[25]

The journalist asked a well-known Chinese scientist to comment
on how the Dominican mine compared with pits in Myanmar. The
answer was damning: 'If one is incredibly safe and 10 is completely
dangerous, the safety of those mines today were [*sic*] a five or six . . .
The mines I've seen in Burma are a nine or 10.'

Located in the Hukawng Valley, in Myanmar's far northern
Kachin State, the Burmese mines are largely inaccessible. This is
due to their general isolation and because foreigners may not visit
the region on account of the contentious political situation. Kachin
State has been riven by conflict for decades. Quite different to Prussia,
there are very few early accounts of mining amber in Kachin. The
earliest written accounts in English are those of the English military
in the 1830s.[26] One, by Captain S. F. Hannay, recorded that the locals
mined amber via a series of pits and holes and that the pits varied
from between 2 and 5 metres (6½ and 16½ ft) deep and were about
1 metre (3¼ ft) wide.[27] Another observed that the pits were slightly
wider and that the walls were cut with holes which functioned as
steps to allow the miners to descend. The tools used were metal crow-
bars and picks, wooden shovels and buckets drawn up by a winch.[28]
Today, pits can be up to 100 metres (328 ft) deep, but they are still
only wide enough for a single person to work in. In particular, the

political situation has made the miners vulnerable to exploitation. Few own the land they dig or the equipment they use. They have no access to free healthcare if injured in the inevitable accidents. To be allowed to dig or to sell, bribes and taxes are paid to representatives of the national government as well as to local Kachin ethnic militias. There is presently huge interest in amber from Myanmar, but several scholars have pointed out that rather than benefitting the miners it finances and perpetuates the instability which mars their lives.

COMMERCIAL OPERATIONS

The first commercial amber-mining operations were in Europe, established in the middle of the nineteenth century. In 1855, works to widen the access to the Curonian Lagoon unearthed copious quantities of amber, prompting interest and investment.[29] In 1861, the Prussian firm Stantien & Becker offered to shoulder the costs and to pay the government daily compensation to continue excavating, if they were allowed to keep the amber.

Stantien & Becker adopted techniques from traditional mineral mining to excavate amber. Their use was facilitated by the recent understanding that amber was consistently being found in a layer of earth noted for its greenish-greyish blue colour.[30] This blue layer typically lay approximately 5 metres (16½ ft) below sea level and somewhere between 40 and 45 metres (131 and 148 ft) beneath the topsoil. In a place called Palmnicken (now Yantarny) the layer was 1–2 metres (3¼–6½ ft) below the surface. Stantien & Becker used diggers to excavate the lagoon bed and the dunes, and to gouge out huge open-cast pits (illus. 42). They initially agreed to make six diggers available for at least thirty days between May and September every year and to pay 30 marks per day. By 1863 twelve diggers were in operation, and Stantien & Becker were paying 45 marks per day. Within five years, they had expanded to sixty days of operation across a greater area, and within a decade it was a year-round enterprise employing 1,000 men. By the end of the century, the firm was excavating more than eight times the amount that could be handpicked from the shallows

42 Amber mine at Palmnicken, East Prussia, late 1890s, photograph reproduced in Emil Treptow, Fritz Wüst and Wilhelm Borchers, *Bergbau und Hüttenwesen* (1900). See also 'Palmnik' atop the dunes in the map of 1677 (illus. 40).

and beaches. A less conventional method used was divers. Friedrich Wilhelm Stantien and Moritz Becker saw the newly invented diving suit at the 1867 Paris exhibition. The equipment and two French instructors were then introduced to the business. Between 1869 and 1885, Stantien & Becker's diving operations employed three hundred men, fifty boats and sixty diving kits. This did not bring huge yields, just under 8 tonnes (17,600 lb) in 1882, and caused considerable damage to the marine habitat.

Stantien & Becker exploited amber on an industrial scale. A new railway brought people, but also coal to power the excavators and the huge hoses which were used to blast amber from the earth.

There was also a railway on site for transferring excavated material from the mine to the processing works. Here amber was blasted out with water, spun in drums to slough off its crust and pre-sorted using sizing machines. Two-fifths were sent for further sorting into ten types by hand. Amber suitable for working into objects and jewellery constituted about a quarter of the complete harvest, and this was subsequently sorted into a further seventy categories. The remaining three-fifths went to be processed into amber oil, amber acid, amber varnish and pressed amber. Hand-sorted amber for which no use could be found was also added to this. In total, about 75 per cent of all amber found in this period was destined for uses other than objects and jewellery.

In 1899, Stantien & Becker's operation was taken over by the State and renamed the Königsberg State Amber Manufactory.[31] Today excavation continues in Palmnicken, which has been known as Yantarny (янтарь, meaning amber) since becoming part of Russia in 1945. Mining continued after the Second World War with the site under national ownership and gave the region approximately 10 per cent of its income. However, the disintegration of the Soviet Union shattered the domestic market and broke links to suppliers and wholesalers, leading the mine to become a public limited company in 1993. This decision was later reversed by the Russian Supreme Court.

AMBER IN THE THIRD MILLENNIUM

For much of the 1990s, and the first decade of the twenty-first century, the local authorities in Kaliningrad pushed national government in Moscow to hand over ownership of the mine and processing plant to the Oblast. But as the mine had been designated an enterprise of strategic importance, this could not be done. This changed in 2011 when the then Russian president, Dmitry Medvedev, finally agreed to exclude the factory from the group of strategic enterprises, reopening the possibility of local involvement.

Today, visitors can stand on the edge of an enormous strip mine at Primorskoje (illus. 43). This was opened in 2002 after the flooding of

43 Open pit at the Primorskoye amber mine, Yantarny, Kaliningrad, Russia, 2015.

the previous mine and exhaustion of other deposits. Up to 60 metres (197 ft) deep in some places, this pit yielded 342 tonnes (more than 75,000 lb) of amber in 2011, a figure well below the glory years of the 1980s when all pits combined produced between 580 and 820 tonnes per year. The mine now operates as the Kaliningrad Amber Combine and heavy investment has led to year-on-year improvements, with a forecast of 500 tonnes from the Primorskoje mine in 2025.

A FRAUGHT FUTURE?

The Amber Combine faces competition from amber traffickers. Their share of the market has grown steadily from 15 per cent in the 1980s to nearly 30 per cent in the new millennium.[32] Recent studies have identified lack of legal regulation and enforcement as pressing problems if unregulated amber extraction and trafficking are to be combatted.[33]

Illegal activities prevent the industry from developing responsibly and sustainably. Attempts to establish legal amber mining in Ukraine have been hindered by unlawful practices. Some 90 per cent of the amber extracted in Ukraine in 2015 was obtained without permission. In 2017, police in just one administrative district were able to confiscate 149 pumps and hydraulic hoses, 58 vehicles and just under 2.5 tonnes (5,512 lb) of amber. It is thought that only 10–15 per cent of the illegally mined amber is ever recovered. The rest leaves the country hidden in vehicle dashboards, bumpers and boots.[34] In Poland, figures suggest that of every 10 tonnes of amber hitting the market only 600 kilograms (1,323 lb) will have been acquired legally, losing the state billions in tax revenue per year. Plans are now afoot to treat mining amber without a government concession as a crime, just as happened in the Prussia of old.[35]

In Myanmar, amber occupies a strange legal position. The principal reason for the current interest in Burmese amber is its rich fossil content. While it has been legal to export burmite since 1995, the export of fossils without permission has been illegal since 1957. It is unclear which law takes precedence. The people of Myanmar are losing their palaeontological heritage at breakneck speed. On the backs of scooters, in cars, boats and even on elephants, huge quantities of burmite are being smuggled across the border to China, where choice specimens command large sums; others are hawked for thousands of dollars, and the rest are transformed into cheap jewellery for sale across the world. Today people are more aware than ever of the environmental devastation that mining, regulated and unregulated, causes. Sadly, deforestation, erosion, loss of habitat and land degradation also go hand in hand with the exploitation of burmite.

Around the Baltic, a community of amateur amber fishermen and women keeps the old techniques and knowledge alive. In collecting amber from the shore and the shallows they leave little physical trace, just as with those who collected amber before the nineteenth century. Is there a market for amber collected in this way? Not yet, although in some Baltic holiday resorts families can pay to be guided in collecting and working amber. There is clearly interest in the collection of amber as a form of 'experience tourism', a trend in which visitors seek to make authentic and once-in-a-lifetime memories. Perhaps born of a type of nostalgia for the preindustrial past, this is far removed from the harsh realities experienced by amber fishermen in Prussia five hundred years ago. It does, however, offer hope for the future and the preservation of knowledge transmitted down the centuries relating not only to the incidence of amber, but local climate, weather and on- and off-shore topographies, all of which are projected to change with global warming.

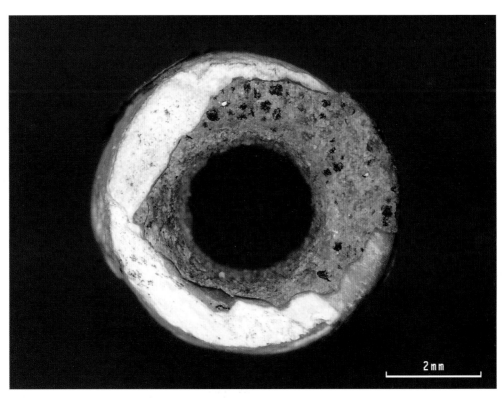

44 Bead made in imitation of amber from Cova del Gegant, Sitges, Spain, excavated from a layer dating to the 2nd millennium BCE.

five

Making and Faking Amber

While amber is found abundantly in some places in the world, it is rare in others. For at least two millennia, indeed longer, rarity, curiosity and cupidity have inspired attempts to make amber imitations. Some of the oldest known hail from Spain and were made four millennia ago (illus. 44),[1] while the earliest recipe for forged amber dates to around 200–250 and comes from China. It instructs:

> Take an egg, mix the yolk and white of it, and boil it. As long as it is soft, an object can be cut out of it; this must be soaked in bitter wine for several nights, until it hardens; then rice-flour is added to it.[2]

Eggs are also the key to the first known European recipe for making imitation amber, which dates to the 1400s.[3] This uses only egg white boiled inside a piece of gut until hard. Shapes can be cut from the mass, greased with oil and left in the sunshine for a week. The author states: 'the more linseed oil you add, the more colour the "amber" will gain; the longer it stays in the sun, the stronger it will be-come.'[4] Amber has been copied, imitated, simulated, forged and faked for centuries, and not only amber itself, but the unique inclusions it contains. The motivations for doing so are complex, and the way in which these creations were and are received is highly variable and sometimes unpredictable. Over time many would attempt to create imitation amber, including renowned figures such as Leonardo da

Vinci.[5] The production of sham ambers and fake flies continues to this day, employing methods of ever-greater sophistication, making modern collectors and connoisseurs as vulnerable as their historical predecessors and generating marvellous stories along the way.

COLOUR AND CLARITY

European recipes for imitation amber focus on recreating the colour and clarity of amber. Often referred to as 'Baltic gold', amber has been prized for its colour for millennia. Pliny the Elder wrote that there were various kinds of amber distinguished by their colours, which he described as pale, waxy and tawny. In his time the last was one of the more valuable 'and still more so if it is transparent, but the colour must not be too fiery: not a fiery glare, but a mere suggestion of it'. He reserved his highest praise for Falernian amber, so-called because its colour resembled that of a wine grown on Mount Falernus, between Rome and Naples, which was white but was aged in amphorae and turned reddish brown. Falernian amber was 'transparent and glow[ed] gently [with] the agreeably mellow tint of honey that has been reduced by boiling'.[6]

Today, it is easy to imagine the glassy tones that Pliny describes. Road users in Britain, the United States, Canada and Australia meet it at every traffic signal. Beer drinkers, particularly of Australian lagers, will know which beer is meant when 'amber nectar' is invoked, although perhaps not that this has a longer history as a descriptor of apple juice. Scots will recognize the description of a haggis's cooking juices as beads of amber and know which author coined the image. Some uses are no longer current. Amber is not now used to describe red hair, as it was in Scottish literature of the nineteenth century and before this in ancient Rome, where Emperor Nero's lover Poppaea's auburn hair was referred to as 'amber'.

Whether in the context of curls, silkworms ready to spin, cider, tea, coconut oil or plums, amber is an adjective which conjures warmth of hue, sheen and beauty. In Europe, amber's clarity and transparency were even proverbial. 'They say of something that is

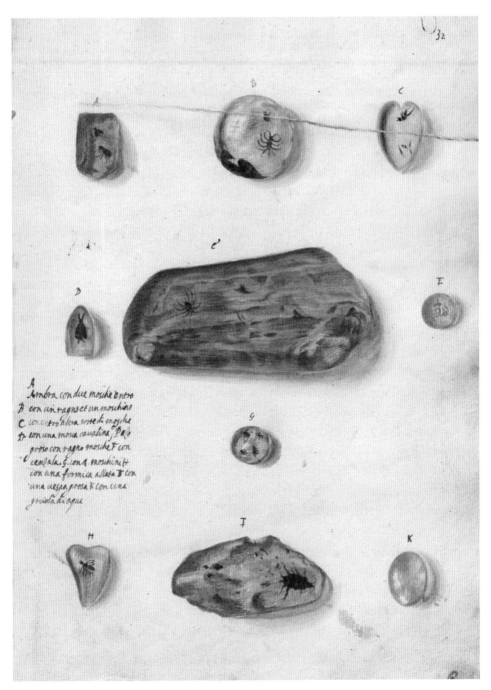

A
Ambra con due mosche entro
B con un ragno et un moschino
C con entro altra sorte di mosche
D con una mosca cavalina, D più
grosso con ragno mosche F con
centola. G con 4 moschini H
con una formica allata I con
una vespa grosa K con una
goiola di aque

50 Watercolour showing inclusions in amber in the *Codice Settala* (1640/1660), an illustrated
inventory of Manfredo Settala's collection preserved in the Biblioteca Ambrosiana, Milan, Italy.

been given to Cardinal Francesco Barberini by Władysław, son of the ruler of Poland-Lithuania, but Władysław had a change of heart.[43]

The routes by which inclusions were acquired became an essential element of the inclusion's own history. Collectors also competed with others to acquire the most unusual inclusions, and tastes were different depending on where one came from. Bees in amber, for example, were especially popular in Italy. There, bees were traditionally said to be divine and were closely related to death. They were also the symbol of the powerful Barberini family, and Pope Urban VIII, a scion of this dynasty, owned a piece with bee inclusions. This was not just a single bee but three together, recalling his family coat of arms.

PUZZLING PRESERVATION

If inventories and accounts are surveyed, it soon becomes clear that there were many more frogs and lizards in amber circulating in sixteenth- and seventeenth-century Europe than there are today. Even in the era of amber's most intense and systematic mechanical excavation, the nineteenth and early twentieth centuries, only one example of Baltic amber encapsulating a lizard is known to have been found. Early modern accounts of the port city of Danzig suggest that lizard inclusions were quite the thing. When the French diplomat Charles Ogier visited in November 1635, he noted that there were frogs and lizards in amber for sale and celebrated them as 'miniature marvels of nature'. Less convinced of their authenticity, the English traveller Fynes Moryson referred to them as 'well prised works of manuall art'.[44] Though amply discussed and even depicted in works of the time, few of Ogier's so-called wonders survive today. To modern eyes, the carved hollows in which the inclusions lie are blatantly obvious (illus. 51).

The gullibility of collectors of amber is a recurring theme in eighteenth-century literature. Johann Christian Kundmann, a member of the Academy of Sciences Leopoldina, recounted the story of one collector in Breslau (now Wrocław):

[He] had a piece of amber in his cabinet of naturalia, in which you could see quite a large frog: he always showed this curious piece himself by holding it up to the light and never let anybody else take it in their hands; . . . I, however, was present at the making of the inventory after his demise and we all saw that this piece of amber had been cut through the middle, then hollowed out, then a frog set in the hollow, and that the pieces had quite clearly been stuck together again.[45]

Friedrich Samuel Bock gave a detailed description of how these 'artful animal burials' were made:

They [the artists] cut rather a thick piece of amber horizontally in order to produce two quite thick plates. Then they make a

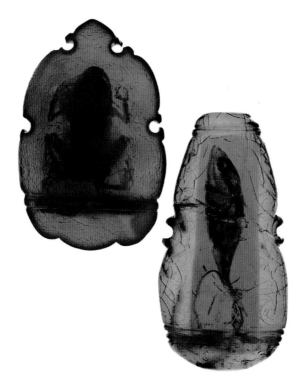

51 Fake amber inclusions collected in the 19th century
by the German natural historian Georg Carl Berendt.

hollow in one or in both according to the size and shape
of the creature that is to be closed up; they lay this into the
hollow and then they fill the crack between the two plates
which have been put together with a mastic glue or with
another amber-like mix. Because the eagle-eyed can spot
the deceit nevertheless, they mount the edges within a
gold or silver ring.[46]

By the eighteenth century, collectors were warned to be suspicious
of frog and lizard inclusions. Tests such as soaking them in warm
water to crack any joins open were suggested. However, debate raged
as to whether they were to be admired on account of the obvious
workmanship and ingenuity involved or abhorred as lies.

PRAISE FOR PRESERVATION

Since Roman times there had been a sense that creatures trapped
in shining golden amber were dignified by it, not least because the
insects found were usually of the dislikeable type: the annoying ant,
the buzzing stinging bee, the scuttling lizard, the slimy frog. The
fact that inclusions were not kept alongside other examples of their
species but stored with artistically worked ambers is evidence that
many were artworks first and scientific specimens second.

Almost all post-medieval accounts of amber and its inclusions
refer to the ancient Roman poet Martial, whose quips on amber-
embalmed creatures are included in his *Epigrams*. He recounted that
ants, usually so despised, were elevated if killed by and entombed in
amber.[47] In the Renaissance, scholars picked up Martial's work in the
service of their own. Giambattista della Porta used Martial to inform
and illustrate his inclusion experiments. He is clearly echoing Martial
when he writes that amber surpasses even Cleopatra's tomb and that
creatures would choose this death above all others.[48] The story of
Cleopatra and Mark Antony was so widely known in this era that
readers would easily have understood which monument was meant
when Francis Bacon wrote of amber that spiders, flies and ants find

'a Death, and Tombe, preserving them better from corruption than a Royall Monument'.[49]

It was not lost on elite collectors of spectacular inclusions that owning them was like owning pieces of poetry. The connection was sometimes even made explicit. The Italian cardinal Scipione Borghese owned an amber chalice, the foot of which incorporated a frog and was inscribed 'latet et lucet' (hide and shine) after the opening of Martial's epigram on a bee in amber.[50]

THE POETRY OF PRESERVATION

Amber's intimate connection with Martial's verse not only led Europeans to catalogue their own inclusions in ditties but to dedicate

52 Title page from the first edition of Daniel Hermann's poem *De rana et lacerta* (1583).

poems to flies, amphibians and reptiles in amber. At some time in the late 1570s, two pieces of amber, each about a hand span in size and each housing a creature, were put up for sale in Danzig. The subject of Marcin Kromer's *De rana et lacerta succino Prussiaco insitis* (On a Frog and Lizard Embedded in Prussian Amber, 1578), they were also celebrated in a sixteen-page ode, Daniel Hermann's *De rana et lacerta* (On a Frog and a Lizard). Today the frontispiece of Hermann's poem is considered the oldest existing visualization of animal inclusions in amber (illus. 52). The pieces appear to have remained in Danzig until at least 1593, when Fynes Moryson saw the 'two polished pieces' there. According to Moryson these 'were esteemed a great price' and had attracted the ruler of Poland-Lithuania to offer a substantial sum for them. Sigismund I lost out to Georg Friedrich of Hohenzollern, Margrave of Brandenburg and Ansbach, and they made their way onwards, probably as gifts, to the Gonzaga in Mantua, Italy.

There have been many literary treatments of inclusions since. Sherlock Holmes mused on 'sham flies' in 'sham amber',[51] and many authors hailing from Prussia, particularly those forced to leave at the end of the Second World War, reference amber inclusions in the titles of their autobiographies. Perhaps the most famous lines, however, are those of English poet Alexander Pope:

> Pretty! In amber to observe the forms
> Of hairs, or straws, or dirt, or grubs, or worms!
> The things we know, are neither rich nor rare,
> But wonder how the devil they got there.[52]

PRACTISING PRESERVATION

Pope's ditty summarizes the key features of people's fascination for amber as amazement and confusion. Amber was highly praised for its ability to give dead creatures the appearance of being alive, so much so that Francis Bacon framed his thoughts on the 'preservation of bodies' around amber.[53] Martial compared the stiffening of amber to the freezing of water, and in Italian the adjective *congelato* (frozen) is often used to describe these creatures. For many centuries ice was the only thing which could preserve soft tissue complete, in three dimensions, unmoving and unchanged for the purposes of study. But ice had its own problems, as it had to be kept cold.

Early natural historians had an inkling that amber was unique in its ability to conserve, but they did not know just how unique. With amber, the processes of fixation, dehydration and sterilization begin as soon as the liquid resin engulfs the inclusion. It hardens so rapidly that it produces a 'hermetically sealed tomb', mummifying the finest of details with virtually no shrinkage and negligible decomposition. These virtues led natural historians to experiment with amber in the conservation of specimens. There was a strong conviction that amber's virtues might be recreated in a laboratory setting and put to practical use. Giambattista della Porta experimented with 'shutting up Things, even forever' and boasted that he had

> made trial hereof in Amber; first reducing it to a convenient
> softness, and then wrapping up in it that which I desired to
> preserve: For whereas the Amber may be seen throw, it doth
> therefore represent unto the eye the perfect semblance of that
> which is within it, as if it were living, and so sheweth it to
> be sound, and without corruption.[54]

Della Porta describes the impossible and he was probably using another amber-like gum or resin. He was one of many experimenting with amber and amber-like masses to preserve specimens. At its simplest, this involved attaching specimens to paper and painting them with an amber varnish. At its most complex and most disturbing, attempts were made to preserve humans, in the belief that such 'transparent tombs would be very proper for persons eminent of their station or beauty'. In one bizarre episode around 1675, Dr Theodor Kerkering of Hamburg reportedly succeeded in enclosing a human foetus in self-made amber while preserving its colour and shape.[55]

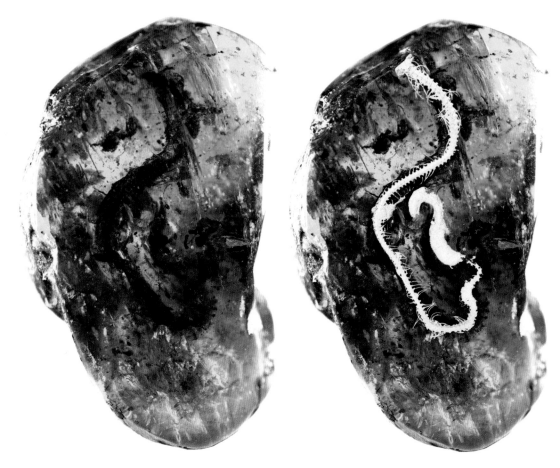

53, 54 Piece of amber with juvenile snake skeleton (left) and synchrotron X-ray micro-computed tomography (CT) image of the same snake skeleton (right).

EXHUMING INCLUSIONS

It was, and is, not only the inclusion but the fusion of inclusion and amber that made, and makes, these pieces interesting and precious. An insect specimen cannot be 'exhumed' from amber by, for example, melting or dissolving away the surrounding amber. Amber requires elevated temperatures and pressures to melt it, which would destroy the inclusions within. In the past, inclusions were sometimes extracted by cutting pieces open, but this too spelled destruction. For years scientists had to be content with readying an inclusion for study only through reshaping, polishing and photography.[56] Though the last can be challenging, it must be said that lighting nuggets from different angles has often revealed hitherto unnoticed fakes by exposing joins or signs of casting, tool-marks and even human hairs. X-ray computed tomography and synchrotron scanning are two of the most promising recent techniques (illus. 53, 54).[57] These innovative methods allow non-destructive digital dissection and capture data which itself can be modelled and printed three-dimensionally.

With time, means of determining true from false will become ever more sophisticated, but then so too will the technologies and materials for synthesizing amber and inclusions. Collectors and researchers the world over vie with one another to make extraordinary discoveries, their competition pushing prices to incredible heights and forgers, unsurprisingly, to unprecedented lengths. With certain single specimens sometimes commanding over £100,000, creative individuals can expect handsome returns if their handiwork convinces and it gives good reason to perfect such shady skills.

55 Bückeburger marriage collar, Germany, *c.* 1850,
amber, silver, base metal, glass beads and textile.

six

Accessorizing with Amber

Today many people associate amber primarily with jewellery. Whether sold in shops or worn by others, bodily ornament is probably the form in which amber is most often met. It is a typical souvenir of trips to Denmark, Poland, Lithuania and Latvia, and it can be easily obtained from a plethora of online boutiques. Amber has been treated as a gemstone and cut, facetted, polished and set for centuries. As with nearly all other gem materials, there have been times when amber has been more or less fashionable, when its design has been more or less innovative, and when the things made from and with it have been more or less affordable. From snuffboxes to powder horns and from swords to parasols, it has also been the material of many other accessories. There are so many that to address all here would be impossible. Throughout history amber has been both a marker of milestones, like the popular northern German collars given and worn to record the changing status of a woman once married, and a must-have fashion detail (illus. 55). This chapter considers the wearing and bearing of amber from a variety of angles – from the maker, marketer, consumer, gifter and wearer – and looks at some of the reasons, many unexpected, for which amber was worn and appreciated.

CONTEMPORARY AMBER ART JEWELLERY

The use of amber for wearable works of art is not well known. Beyond its traditional Baltic heartland, amber has not been popular among

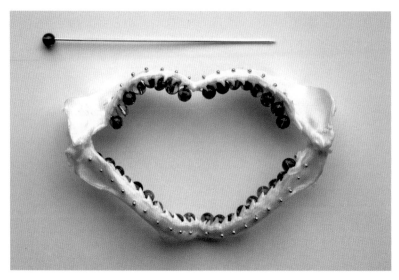

56 Herman Hermsen, 'AlaDali' brooch, 2015, shark's jaw, amber beads and gold.

artists creating their own jewellery or collaborating with designer-craftspeople to do so. Around 90 per cent of Polish amber production involves commercial, mass-produced, souvenir-type jewellery. For more than a decade, Sławomir Fijałkowski, head of the Experimental Design Studio at the Academy of Fine Arts in Gdańsk, has advocated greater provocation, risk-taking, innovation and originality in Polish amber jewellery. This goes hand in hand with the awarding of the Amberif Design Award, a prize which engages amber jewellery in global design conversations.

Both Fijałkowski and Amberif are focused on a modern identity for amber and greater popularity among sophisticated young audiences. Two established artists who have worked with amber throughout their careers are Herman Hermsen (illus. 56) and Heide-Marie Herb. They engage and experiment with amber in ways which overturn conventions and clichés. In their hands, gleaming beads of amber become the teeth lining a menacing shark's jaw or geometric amber cubes and prisms are combined with precious metal mounts echoing girders and other industrial armatures. At the other end of the spectrum, visual artist Yu Ji exaggerates and explodes amber necklaces to unwearable dimensions. In her *Etudes-Lento IV*, rusty metallic chains hang in a tangle from the gallery ceiling, each thick with an oozing and dripping, amber-coloured goo.

THE CREATION OF TRADITION

The works of Hermsen and Herb challenge the long enduring perception that amber is the domain of grandmothers and great-grandmothers. For many people, amber is an heirloom. It was last very fashionable in the 1930s and '40s, and it was most fashionable in Germany. The Roman author Tacitus had singled out amber in his discussion of the peoples of Germania.[1] This led to amber being

57 Cover of an English-language marketing brochure for the Staatliche Bernstein-Manufaktur (State Amber Manufactory) Königsberg, with a portrait of Daisy, Baroness Freyberg zu Eisenberg (Daisy d'Ora), *c.* 1930.

promoted by Germany's National Socialist government.[2] Germans and the German diaspora were encouraged to embrace amber as symbolic of their history.[3] Shop windows displayed amber jewellery beneath figures in 'Germanic', Baroque and modern dress and the slogan: 'For Centuries a German National Treasure'.[4] Discussed in terms of its blondness and compared to 'German curls' and 'ripe sheaves of corn', it was presented as the perfect accessory for patriotic women and modelled by socialites and stars (illus. 57).[5] Exhibitions of Königsberg-made ambers (now Kaliningrad) toured Germany and its annexed territories. Souvenirs of amber were made for the Berlin Olympics in 1936, and there were even luxury bindings of amber for Adolf Hitler's *Mein Kampf*.[6]

Perhaps the most interesting of these phenomena were the more than 60 million tiny pieces worked into lapel pins and pendants and given in return for a donation to national welfare charities. Their shape changed frequently, sometimes representing oak and clover leaves, sometimes spring flowers, to encourage people to give regularly and to establish collections, guided therein by a specially produced handbook. One museum curator commented that, though each badge seemed 'little more than an esoteric symbol', each had their own value as a sign of 'hope that we will be helped and pride that we are able to help'. He emphasized that 'only amber could express this so clearly' because 'an inner need of the soul' drew Germans to wear it. Writers suggested that Germans had the duty to buy their national stone, and some were even of the belief that amber should only be worn by Germans.[7] One rhymed:

> German stone from German soil
> Hold it proudly in your hand
> You are asked to carry this
> Because you love your Fatherland.
> The ancient elemental forces, which have shaped you
> Simple German gem
> Radiate and proclaim:
> 'My bearer should be German too.'[8]

BOOM AND NEW BUST

The European amber industry had been facing challenges for quite some time. In the nineteenth century it had grown to an enormous size with a global clientele. The volume and variety of raw amber available had expanded immensely since the 1850s thanks to diving, dredging and water-blasting. Used for necklaces, bracelets, hair slides and the like, jewellery-grade amber was only one type. Something like a quarter of all jewellery-grade amber remained in Germany, and the rest was sent across Europe, as well as to Turkey, Mexico, India and Hong Kong. Each country had its own market, fashions and traditions. In late Victorian England, the Aesthetic movement encouraged women to wear amber as a relaxed and natural material.[9] In the Ottoman Empire and across the Islamic world, amber beads were being used to make *misbaha*, strings of prayer beads for the saying of *tesbih*, and to punctuate the chains from which *hirz*, amuletic Koran boxes, were suspended. Within the Moroccan Jewish community, beads were believed to bestow longevity and were used in women's dress, and in Iraqi Kurdistan they were used in the suspension of amulets and other pendants.

From around 1900 the extraction of amber began to be increasingly unprofitable. The Versailles Treaty, following the First World War, caused further problems. West Prussia with Danzig (Gdańsk) was ceded to Poland. East Prussia with Königsberg became a semi-autonomous state protected by the League of Nations. Its isolation from Germany proper meant poor routes to market; German goods were unpopular in the post-war years;[10] and amber itself faced competition from cheaper visually similar synthetic materials.[11] The National Socialist focus on amber rescued the ailing industry from collapse.

AMBER FOR SMOKING AND SNUFF-TAKING

Obtaining and processing amber created enormous amounts of waste. It was discovered that fragments could be heated and fused to make a mouldable, dyeable mass. This was especially suitable for making mouthpieces for pipes, cigars and cigarettes (illus. 58). In Vienna,

58 Cigar holder showing the coronation of King Ludwig II
of Bavaria, 1864–7, amber and sepiolite.

where pressed amber was patented, it was typically combined with
wonderfully carved sepiolite pipe bowls or sometimes embellished
with enamels, as designs for cheroots by the artists of the Wiener
Werkstätte show.[12] Further east, in the Balkans, pressed amber mouthpieces were combined with filigree silver, and further east still, in
the Ottoman Empire, pressed amber hookah mouthpieces were set
with gems.

Modish amber mouthpieces were the culmination of a long
association between amber and tobacco, specifically snuff – a spiced
ground tobacco. Nearly two centuries before, amber had been used for
pocket-sized snuff containers, which, in the later seventeenth and into
the early eighteenth centuries, were indispensable gentlemen's accessories. Though many were made by craftspeople in Prussia, snuffboxes
were also the work of specialists in major metropolises. The tri-lingual
trade card of one London goldsmith advertises his ability to make a
'great variety of curious work' in amber and advertises his willingness
to buy 'amber, jewels, and curiosities', perhaps for transformation into
other equally fashionable pocket pieces like watch cases.[13]

In the early eighteenth century, the European practice of snuff-taking and the use of dedicated containers to store it inspired a fashion at the Qianlong imperial court in China. It did not take long for Chinese craftspeople to begin making snuff containers to meet the tastes and needs of a local clientele. Snuff bottles were a perfect use for pebble-shaped, fist-sized pieces of amber. The bottles are notable for the variety of skills they display. Bottle-shaped flasks are carved so thinly that light passes through them (illus. 59). This is no mean feat when one considers the long delicate process of hollowing out the interior and reducing the walls through a tiny opening. Bottles shaped like gourds and melons cleverly exploit amber's textures, mottled tones and flashes of colour. The first study

59 Snuff bottle carved with women and children in a garden,
Qing dynasty, Qianlong period (1736–95), amber, ivory and coral.

of Chinese snuff-taking, *Yonglu Xianjie* (Researches Done during Spare Time into the Realm of Yonglu, God of the Nose) specifically highlights the appeal of amber's variegated tones. Painted details were sometimes added for extra interest, and many bottles were also engraved, particularly with poems and messages of luck. Chinese snuff bottles became popular collectors' items in the nineteenth century, especially among Europeans, and demand continues to be strong in Europe, Asia and the United States today.

BATTLEGROUND PRESSED AMBER

European commentators on amber at the end of the nineteenth century were not positive in their assessment of fused amber. Eugen von Czihak, director of the School for Civil Engineering, Königsberg, called for amber to be treated with greater respect.[14] He was inspired by a group of local producer-protestors: the Association for the Utilization of Amber in the Applied Arts.[15] They argued that pressed amber devalued the genuine article through its use to make 'miserable products', such as

> mirror frames strung together from rose petals and flowers, senseless knick-knacks, obelisks for thermometers, clock cases in the form of clumsy sideboards, inkwells in the shape of guardhouses atop squares cobbled with small amber tiles and fenced with chains, and . . . ugly . . . bluffs of amber crawled over by metal lizards and frogs.[16]

The Association also found contemporary jewellery lacking and felt that commercial jewellery did only 'modest justice to this multi-coloured material with its matt shine'.[17] The German curator Otto Pelka, writing slightly later, was less diplomatic. He damned the later nineteenth century as an 'era of . . . brooches, bracelets and necklaces' made by artisans who had 'increasingly limited themselves to turned and carved mass-produced wares with no claim to artistic value'.[18] The Association called for amber to be used in ways appropriate to its nature, texture, colours and properties. They sponsored a display at the 1900 Exposition Universelle in Paris.[19] Presented within amber-studded walls and centred on an amber-encrusted column, the display showcased, among other things, a salver decorated with amber marine life and furniture inlaid with amber intarsia.

International fairs were places to show that Baltic amber could adapt to modern needs: amber-headed buttons for electric doorbells were just one suggestion. Prussian jewellery, however, struggled to rise to the challenge. Discerning customers had to turn to Denmark where Georg Jensen and Bernhard Hertz made buckles, brooches and pendants combining amber, silver and green agate (illus. 60); to the Netherlands where Van Kempen & Son and Van den Eersten & Hofmeijer used amber in their vessels and jewellery; to Vienna where the artists of the Wiener Werkstätte used amber in hat pins

60 Georg Jensen, Master Brooch no. 96, Copenhagen,
Denmark, *c.* 1900–1910, silver, amber and green agate.

61 Vanity and cigarette case, Cartier, Paris, *c.* 1920, enamel, diamonds, chrysoprase, amber, mirrored glass, gold and platinum.

and brooches; and to Paris where Christian Fjerdingstad employed amber in his designs for Cristofle, and where Cartier created luxurious vanity and cigarette cases combining black enamel, amber, jadeite and diamonds (illus. 61). In Prussia, the first stirrings of design-led jewellery were in the 1930s, when the state-sponsored Amber Manufactory in Königsberg (Staatliche Bernsteinmanufaktur) invested in the appointment of artist-goldsmiths such as Toni Koy, Jan Holschuh and Hermann Brachert.[20] Their early works are individual responses to the beauty of each nugget, but as employees of the Bernsteinmanufaktur their production also manifested the attitudes and aspirations of the National Socialist German Workers' Party.

BEAD-MAKING

For authors such as von Czihak and Pelka, jewellery was the perfect subject around which to base any discussion of amber's past, present and future. After all, no other use for amber had as long a history as bead-making. Today the basic steps are much the same as those followed in medieval Europe.

century. Devotees focused their prayers on events from the life of Jesus Christ and his mother Mary and used a string of beads to keep their place in these prayers.[37] The prayer and the beads used to say it are known as the rosary. Saying one Our Father, ten Aves and a Gloria needs a minimum of eleven beads if the Our Father and Gloria are said over the same bead. The pattern had to be repeated five times, once for each of the five mysteries of the Virgin, meaning that strings might number fifty beads or more (illus. 63).

BALTIC AMBER HEADS EAST

There was also a considerable market for amber beads beyond Europe. Amber's popularity in the Near and Middle East surprised travellers and baffled Prussians. From the mid-sixteenth century onwards, Armenian and Jewish diasporic networks facilitated the arrival of amber in western and central Asia via a myriad of land and river routes. In time, Armenian traders would become so intimately associated with it that, by the beginning of the eighteenth century, people carrying long strings of amber beads in London were simply assumed to be Armenian. In Constantinople (modern Istanbul), Western European visitors marvelled at the wealth that amber merchants amassed. One Danzig merchant even went there as 'he wondered what use the Turks put so great a quantity of it [to].'[38] It seems to have been found across the whole region. Sixteenth- and seventeenth-century travellers explained that it was used to trim 'reins, the saddles, and the saddle blankets' of horses and in the decoration of camels.[39] Pietro della Valle saw Bedouins near Baghdad in amber necklaces, perhaps in combination with silver beads, discs and coins as is still typical of Yemeni and North African jewellery, and he saw locals tie smooth and facetted amber beads around their ankles with stalks of straw in Shiraz.[40] By the 1690s, it was claimed that the market 'in Austria, in Germany, in Poland and around Venice' was nothing in comparison.[41] These sources do not mention prayer and worry beads, the uses most seen today in western Asia. The popularity of amber for objects held in the hands means that the awareness that amber has a scent, warms to the touch and becomes statically charged is widespread in the Near and Middle East and almost unknown in Europe where amber is customarily worn rather than handled.

Amber is also particularly well known across the southeastern Europe and western Asia for its historical use as an accessory to smoking, primarily in mouthpieces for hookah pipes (illus. 64). These built on a long tradition of utilizing amber for items to be placed in the mouth. In the seventeenth century, Christian Europeans understood the use of amber for implements such as spoons to result from Islamic

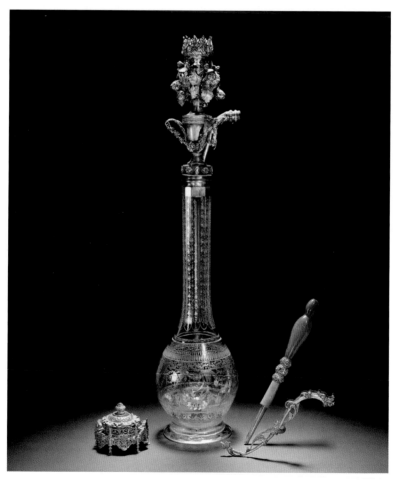

64 Hookah base with mouthpiece, tongs and coal tray, Bohemia
and Ottoman Empire, 19th century, glass, silver and amber.

prohibition of gold and silver objects for eating and drinking. The nineteenth-century French author Théophile Gautier visited Istanbul and commented:

> at Constantinople where [amber] is very dear, the Turks
> prefer it of a pale lemon colour, partly opaque, and desire
> that it should have neither spot nor flaw, nor vein; conditions
> somewhat difficult to combine, and which greatly enhance the
> price of the mouthpieces. A perfect pair of them command as
> much as eight or ten thousand piastres . . . a collection of pipes
> worth 150,000 francs is not at all an unusual thing among the
> high dignitaries or the richer private persons, in Istanbul . . .
> it is, in fact, an Oriental mode of displaying the possession
> of wealth . . . no Turk, who has any self-respect, uses anything
> but pure amber.[42]

The pipe, like jewellery, had a close relationship with a person's face and hands, and was an important and visible social marker.

BEADS FOR BAUBLES

There is a shift in the use of amber beads in sixteenth-century Europe. The discouragement of the use of rosaries in areas that had become Protestant saw beads lose their prayer function and be worn without any religious reason. According to one chronicler, certainly exaggerating, amber was more popular than gold. The wearing of amber was generational. Older women sold their ambers because they considered it unseemly to 'wear brightly coloured necklaces or jewels'.[43] Amber necklaces and bracelets were largely worn by younger women and by children (of both sexes), the latter probably because amber was thought to protect against bad spirits.[44] By the 1690s amber was employed 'for breeding teeth easy', a use which controversially continues today.[45]

Surviving European neckpieces are fairly simple affairs when compared with the official necklaces worn by court officials in

65 Necklace (*chaozhu*) with its original wooden box, 19th century,
jadeite, amber, rose quartz, silver, coral and silk.

contemporary China. The *chaozhu* was a long rope of more than a
hundred beads (illus. 65). These necklaces developed out of the much
older tradition of chanting Buddhist sutras using amber counting
beads, and became popular in China after the gift of such a rosary
by the Dalai Lama in 1643. Dynastic laws and ceremonial sumptuary
regulations dictated and enforced their wearing. These regulations
stated, for example, that the Emperor was to wear yellow amber beads
when visiting the Temple of the Earth. Yellow was not only the colour
of the earth (centre) according to the Qing (1644–1911) concept of
the Five Colours of the Universe, it was reserved for imperial use and
was worn by imperial concubines.[46] The Jesuit missionary Alvaro
Semedo understood the amber beads to have come from Yunnan, in
western China, and he explained they could be used to treat mucus
in the nose, throat or sinuses.[47]

Western visitors to China also wrote home about the single beads of amber they saw tied to the crests of court officials' caps, but they said nothing about the many surviving carved beads and pendants suspended from objects such as fans, not unlike phone charms today. In both Asia and Europe, it was common to use amber beads to decorate textiles, not least because light amber barely added to the weight of heavy velvets and layered silks. In Europe, tubular and barrel-shaped beads were used to trim garments and this may be why amber's sale was sometimes restricted to mercers. Amber beads were used to stud girdles and hairnets in both medieval and Renaissance Europe and in Ming China.[48] In late medieval England,

66 Timepiece in the form of a poppy flower's seed capsule, early 17th century, amber, silver and brass.

amber was given 'stalkes and headdes of goulde' and used as buttons, the most luxurious of which were encased in gold cages.[49] Much later there would even be luxurious, wearable timepieces (illus. 66). The continuation of this practice can be seen in twentieth-century fashion plates showing the judicious use of amber beads to punctuate the drape and fold of Edwardian clothing.

AMBER'S SMELL

In some early modern written sources it can be difficult to tell amber from ambergris, a highly valued substance from the digestive tract of the sperm whale.[50] This is because the word 'amber' can mean both in many European languages. Payments for small amber beads in court embroiderers' account books might refer to amber or to ambergris, and garments described as embellished with amber could have used one or the other. The confusion between amber and ambergris arises, in part, because both were valued for their scent. Amber's smell was part of its appeal in medieval and Renaissance Europe. Many authors recommended amber in 'place of incense' and lesser-quality amber was put to this use.[51] The fact that amber can be burnt is reflected in the name given to it in most northern European languages, literally 'burning stone'.[52] Amber was also widely recommended as an air freshener. It was mixed with petals and spices to make potpourri or cast onto glowing irons and coals. This released a dense dark smoke which was believed to be purifying, so much so that amber was used in mass communal fumigation in plague-ridden London.[53] Miraculous powers were also sometimes attributed to its smoke. One doctor claimed to have seen the instantaneous revival of a patient after fumigation.[54] Some said that, burned wet, it could exorcize malevolent spirits. Today, beliefs in amber's health benefits continue in its use to build saunas. Some 3 tonnes (6,614 lb) were used for the spa at Atostogų Parkas, Lithuania.

Grinding amber to powder was a further way to release its smell. Mixed with oils and waters, powdered amber was used to perfume gloves and scent the skin: two ways of 'wearing' amber that are seldom considered today. Distilling amber produced amber oil. Johannes Magenbuch, the doctor who discovered this, announced it to the duke of Prussia via a gift of amber oil, amber-water and amber-infused bonbons. Many royals in this era were interested in scientific experimentation and appealed to the duke for amber to use in their own experiments. Later printed recipe books shared methods for producing amber oil with a wider audience. They tell readers to dissolve powdered amber in warm linseed oil or alcohol. The latter was less greasy and therefore suitable for garments. Giuseppe Donzelli told his audience to mix a pound of pulverized amber with an equal quantity of wine and to heat the mixture until a golden dew distilled. He warned readers to be careful lest a 'red, tenacious and stinking' oil result. Dark oils would have coloured fabrics, and Donzelli also proposed a recipe for salt-bleaching amber oil.[55] Though sources relating to the Dutch East India Company suggest that there was no market for amber oil in the Far East, there is some anecdotal evidence that the peoples of central Asia used it in the dyeing of animal skins. In England, a mixture of amber, yellow ochre and mastic was also recommended for dyeing leather yellow.[56] Fascinatingly, today amber is used in the production of textiles. Scientists have exploited its astringency, which causes tissues to contract and bleeding to reduce, and processed it into surgical fibres.[57]

ACCESSORIES AND WELL-BEING

Amber oil was highly desirable in early modern Europe because of a conviction that amber's smell was 'effective in the encouragement and maintenance of health [and] in hindrance and destruction of all illness'.[58] Whether in Germany, Italy, France or England, there were many ways of benefitting from amber's smell, the most conspicuous of which – after using amber-handled riding whips and powdering one's hair (later wigs) with crushed amber – was the use of a pomander. Ancient philosophers and physicians had written of the 'spiritus animalis' (psychic spirit): something produced in the brain and filtered or distilled from 'vital spirit' absorbed from the atmosphere. The nose

67 Beads with pomander from the estate of Magdalena Sibylla of Brandenburg,
probably Königsberg, early 17th century, amber and silk.

was thought to be a direct route to the brain, and amber's smell was
held to directly support the production of 'spiritus animalis'.

The name pomander comes from *pomum d'ambra* (apple of
amber/ambergris). Palm-sized chunks of amber were the simplest
pomanders. These just had to be rubbed to release their fragrance.
Perhaps more familiar are the globe-shaped jewellery-type pomanders which were worn suspended from the neck or waist.[59] Some are

made of metal and segmented like an orange. The segments held aromatics, and surviving examples inscribed 'Bernstein' (German for amber) confirm that amber had a place in them. Others are pierced spheres and look like modern tea infusers. These were filled with scented pastes and were sometimes even made of amber themselves (illus. 67). There was also something called a vinaigrette, a little scent bottle with a pierced stopper that could be filled and refilled with amber oil; again some are even made of amber. Amber oil could also be used for improvised pomanders; sponges could be soaked in it or it could be allowed to congeal in 'little hollow boxes of Cyprus wood, of juniper or of bay'.[60] But it could also be used alone. Epilepsy and depression were treated by smearing amber oil under the nostrils. Dizziness, seizures and tremors could be treated with amber oil applied to the temples.

In China, amber was appreciated as an aromatic as early as the Song dynasty (960–1279). At the time of the Liao dynasty (916–1125), which reigned in present-day northern China and Mongolia, amber was itself used to produce the containers for spices, teas, medicines and cosmetics. Chinese sources dating back to 200 CE describe the use of amber for neck and head supports (called pillows), and one much later text, dated to the Song era, recounts the story of one emperor who accepted just such a pillow, not because of its use during rest and sleep, but because it could be sacrificed for medicine.[61]

AMBER PENDANTS AND PORTRAIT PENDANTS

Since no Chinese amber pillows survive, it is impossible to know anything of their appearance. They may have been intricately carved, like other surviving amber accessories which are now treated as artworks. Surviving artefacts from Liao burials, both carved containers and purely decorative pendants, exhibit incredible imagination in the transformation of amorphous amber nuggets into plastic forms: compact river boats with crew, writhing dragons, nesting geese, cicadas and feathered phoenixes. Chinese objects share an attention to amber's material characteristics and respond sensitively to the

72 Pair of wedding knives inscribed ANNA MICKLETHWAIT ANNO
1638, steel, amber and ivory.

73 *Jambiya* dagger, probably Morocco, 19th or early 20th century, steel, amber and coral.

eighteenth-century adoption of amber for the hilts and pommels of ceremonial swords. Swords were an important part of men's costume and their grips opportunities for stylish self-expression or shows of status and wealth. Just as the sword had been, walking canes would become an essential part of a gentleman's daily wardrobe and amber was unsurprisingly also used for their handles, as well as later for ladies' parasol handles.[104] Though the use of walking canes was reserved for those above a certain standing, this usage was open to a wider demographic than swords had previously been, leading to some unexpected insights. In 1708, one man wrote to the Royal Society about a bizarre experience had at home. His amber cane handle had emitted crackles and sparks.[105] This crackling was the result of something called triboelectricity, a phenomenon whereby materials become electrically charged through friction, as happens when a plastic comb is run through hair. Triboelectricity was why amber became 'magnetic' when rubbed and attracted light particles. This 'power', which was primarily observed through amber, led to the Greek word ἤλεκτρον (elektron) being used to describe it, giving us the noun electricity today.[106] Amber's 'magnetism' became a powerful metaphor. In 1711, Lieutenant-General Sir Francis Nicholson, a British soldier and colonial official, met a representative of the Five Nations in New York and presented him with an amber-headed cane 'in memory of himselfe and in token that as ye said head when warme is of an attractive power, so his and their loves should be warme and attractive to draw each to other'.[107]

The amber handle must have seemed a marvel to the chief, whose name is unrecorded. The indigenous peoples of North America had been using amber long before the arrival of Europeans. Around 11,000 years ago, some of America's very earliest inhabitants used amber-based adhesives to fix stone spear points to shafts, and the Thule ancestors of the Inuit turned locally sourced amber into beads. Many early European colonists encountered and wrote about amber in places such as the Carolinas. Yet compared with the Baltic amber field these deposits are small, and so too are the pieces found. As in Africa and India, Europeans seeking the upper hand leveraged Baltic amber. In America, however, the archaeological record has yet to suggest that Native Americans were beguiled by the yellow stone.

MORE THAN JUST JEWELLERY

Down the centuries, accessorizing with amber has been about more than just jewellery. This chapter has covered uses as diverse as dyeing and scenting gloves, prayer aids, reminders of friendship and deadly weapons. Amber was a truly global commodity which, sold rough or in simply worked beads, was versatile enough to allow consumers to engage with it in uniquely local ways.

Of all the accessories discussed in this chapter, only jewellery and perfume are still going strong. Should anybody today want to scent themselves or their homes with amber, the 'golden stone' is still a commonly cited ingredient of many perfumes and home fragrances. There has also been a revival of amber's use for knick-knacks such as carriage- and desk-clocks, trinket boxes, ornaments, picture frames, artistic panels, religious icons and fridge magnets. The clothing in-dustry has not been motivated by amber in the same way as by other precious materials. Gold and silver have inspired lamé fabrics and sequins. Diamonds have spawned rhinestones. Nonetheless, amber's characteristic colour and transparency have still prompted creative responses. In 2019, the Italian lingerie-maker La Perla launched its

'Ambra' range with garments distinguished by a mustard-yellow tone and translucency. Perhaps the next thing will be a revival of the use of amber for lenses, something popular in eighteenth-century Germany and known from nineteenth-century China. Amber-tinted lenses block the blue-wavelength light emitted by the screens of laptops and smartphones that is associated with insomnia in their users.[108] Could spectacles be the future of wearable amber?

The earliest known accounts of amber being used to create altarpieces date to the 1570s. By this time, the Teutonic Order had lost
its hold over Prussia following the conversion of its Grand Master
to Protestantism. As the rulers of Prussia had always done, Albrecht
of Hohenzollern, who became duke of Prussia on his acceptance of
Lutheranism, commissioned works in amber. Initially he had to place
orders in Danzig, where there had been a guild of amber workers
since 1480.[14] The Order had long banned the general possession and
working of amber in their territories, and Albrecht did not have a
court amber-carver. The first known carver, Stenzel Schmidt, was
appointed to undertake 'all work . . . be it small or large carving' in
June 1563.[15]

Albrecht also leased the rights to amber to the Jaski – the family
of well-connected merchants, who, in the words of one of Albrecht's
confidantes, opened the doors to markets in 'Italy, France, Spain,
Turkey and the heathen lands'.[16] By all accounts, the Jaski pioneered
the idea that the type of amber traditionally reserved for bead-making
(known as turning-stone) could also be used for 'fancies . . . such as
spoons and little salt cellars'.[17] This turn in production appalled the
papal legate to the region, Giovanni Francesco Commendone, for
whom the coming of Lutheranism to northern Europe was reflected
in the abandonment of tradition:

> They have stopped making figures of Jesus Christ and of the
> Saints, which pious people once paid so much for. They no
> longer make such great numbers of chaplets or rosaries which
> women used to use for their prayers and ornament too, amber
> being a material of luxury and of piety in one.

According to Commendone, the Danzig masters, then numbering
around forty,

> no longer employ[ed] this precious material for uses other
> than the profane; and they no longer work[ed] it other than to
> make chess- and draughts-pieces, spoons, a thousand kinds of

little vases, and birdcages, all turned very agreeably, but with no practical use thanks to their fragility.[18]

The types of objects Commendone complained about did in fact exist. Duke Albrecht, for example, sent amber spoons to the famous Protestant reformers Martin Luther and Philipp Melanchthon. Luther suffered from kidney stones. He and many others requested amber from Albrecht to be used in treatments. Using an amber spoon may have been an elegant way to enjoy amber's supposed health benefits while highlighting an association with the duke.[19] Prussia was the first territory to convert to Protestantism, and thanks to Albrecht's encouragement of a strong Prussian identity for amber, not to mention gifts to leading Reformers, the material was also connected with Lutheranism.[20]

However, amber spoons were clearly also impractical. Many inventories record broken examples. Comic authors mocked those who used them: Sir Epicure Mammon, the character symbolic of gullibility, greed and vice in Ben Jonson's *The Alchemist* (1610), is derided for his dreams of one day being wealthy enough to own 'spoons of amber headed with diamond and carbuncle'. It was exactly this ostentatious use of an inherently unsuitable material which made them suitable gifts to rulers and royalty: few are likely to have been used for dining and they were probably objects for show instead. In London, Anna of Denmark had four, probably sent to her by the then duke and duchess of Prussia, who are specifically recorded as selecting amber for spoons. The document does not go into detail, but pieces had to be large enough to create a shallow bowl. At least one sixteenth-century source explains that chunks of amber as large as a human head needed to be used if goblets were to be created.

Surviving amber goblets, plates, cutlery, basins and ewers are innovative uses of amber, but stylistically they are related to fashionable shapes already well established in silver- and glassware used in elite dining. Goblets are the most numerous survivors. The earliest securely dated extant example is that recorded among the possessions of the heir to the Italian duchy of Parma in 1587.[21] It is

79 Two plates (from a set of eighteen) decorated with the coat of arms of Brandenburg and Lüneburg and the initials of the text 'Sophia Markgräfin zu Brandenburg geborne Hertzoginn zu Braunschweig und Lüneburg', Königsberg, 1585, silver, silver-gilt and amber.

inlaid with images from the Bible, little portraits of the Evangelists and their symbolic animals. Though this imagery suggests a sacred function, the goblet lacks a golden or gilded bowl and therefore could not be used to say Mass. Other surviving goblets exchange these details for portraits of rulers and coats of arms. Together, they suggest that whoever was making them mastered a series of standard forms which could subsequently be customized and which were to be appreciated for their beauty and skill rather than practical function. Some were made to be quite flashy. A goblet in the possession of the dukes of Bavaria was studded with sparkling glass-paste gems imitating emeralds and rubies.[22] On the one hand, this approach to the material closely resembles the Chinese manner of its appreciation. Surviving Ming and Qing amber cups for wine-drinking and libations are also carved from single chunks of amber, often in ways which clearly echo forms already established for jade. On the other, the approach differs significantly, since Chinese amber vessels, whether cups, brush-washers or other utensils such as pen-stands, are made uniquely and exclusively of amber and eschew the introduction of other materials. Artisans there were trained to create visual interest by paying close attention to the qualities, tones and textures of the original 'stone', drawing them out through the clever use of form.

Letters suggest that some of the goblets at European courts were certainly gifts from Prussia. The 'drinking vessel with case of solid amber' owned by the queen of France at the time of her death in 1592 had come from the then duchess.[23] This could have been Marie Eleonore, Duke Albrecht Friedrich's consort, or Sophia, wife of Georg Friedrich who governed on behalf of Albrecht Friedrich. Duchess was also the title given to Albrecht Friedrich's daughters. His fourth daughter, Eleonore, sought out amber to give to others. Not all of Eleanore's gifts were special commissions. 'The best that one can get at short-notice' sometimes had to suffice.[24] Sophia, for her part, is thought be behind a remarkable set of eighteen amber-and-silver plates bearing her initials and the date 1585 (illus. 79).[25] They are the first of a handful of surviving salvers and basins. All use as much amber as metal and are the result of cross-disciplinary collaboration. Sophia's plates are stamped with the mark of Andreas Kniffel, then court silversmith in Königsberg. Stenzel Schmidt probably turned their wells. He continued to work for the court after Albrecht's death, before being succeeded by Hans Klingenberger. Hans demonstrated a broad range of skills, receiving commissions for objects as diverse as caskets and riding-whip handles. So great was his workload that he sometimes had to partner with others.[26]

PLAYING WITH AMBER

It was probably the production of games-boards that kept Klingenberger so busy. Because Baltic amber was reddish or yellowish, clear or cloudy, it was well suited to making the contrasting playing pieces. Nubs of amber had been used for counters and dice

80 Games-board, Königsberg, *c.* 1608–47, amber, ivory, wood and metal.

by the Romans and Vikings, but the amber-clad games-board was a recent innovation. The earliest known example, now lost, was commissioned from the amber-turners of Danzig by Dorothea, Duke Albrecht's wife, for her brother, Christian III of Denmark.[27]

Making a board was hugely different to turning spoon, goblet or plate. Board surfaces needed a craftsperson capable of sawing out identically sized thin slivers of amber (illus. 80). Their delicacy was key to the successful appearance of the final object, as may be seen in the earliest surviving board. The slivers have been painted on the back. When given an underlay of metal foil, the design can be read through the amber. On one of the oldest surviving board, the effect has been used to create panels which display, among other things, the arms of Landgrave Moritz of Hesse-Kassel, to whom it belonged, and the date. Another board once in Moritz's possession is coffered. It uses the same technique and is signed and dated HK 1611 (Hans Klingenberg 1611).[28]

Cardinal Commendone was especially against objects which had no didactic purpose, and games-boards were something in between. The coffered board is decorated with Latin maxims plucked from the *Sententiae*, moral sayings in verse collected by Publilius Syrus. The sayings cited emphasize the game's relevance to real life. It is certain, however, that Commendone would have been extremely disappointed if he had known that many of the frivolities he abhorred were as admired in Roman Catholic Italy as they were in the Protestant north. The ducal collections in Mantua were home to an early example of the 'thousand kinds of vases' he had criticized. Inventories describe these as being urn-shaped or eight-sided, and some were even mounted in gold and lavishly decorated with gems. Amber birdcages in Dresden and Munich were matched by an example in Florence, where the Medici grand dukes had their collections.[29]

AMBER IN FLORENCE

In October 1587, Grand Duke Francesco died, prompting a sweeping evaluation of his possessions, including his collection. The documents this generated show that amber was at home in many contexts at his court.[30] The grandest was the ducal *Tribuna*, an imposing octagonal room at the heart of the Uffizi. It was here that people could see the amber birdcage, alongside an amber cup and a lizard in amber, all stored atop, hanging from or within the shelf set with drawers ringing the room at eye height.[31]

Yet it was the duchesses of Florence who were best known for their passion for amber. The exceptionally pious Archduchess Maria Maddalena of Austria, Grand Duchess of Tuscany, amassed an enormous collection of devotional ambers, which she kept in her private chapel. The effect must have been breathtaking, particularly when lit by flickering candles, perhaps themselves in amber candlesticks. People seeking Maria Maddalena's favour played to her love for it. Philipp Hainhofer suggested amber as a gift from his patron the duke of Pomerania-Stettin. Hainhofer had been offered an amber crucifix by a contact in Danzig, and he thought that making it up to an altar set by adding a chalice, bowl, ciborium and candlesticks would allow a fair price to be agreed. He floated the idea by Maria Maddalena, who was initially keen, but changed her mind completely when she heard that her sister, the wife of the ruler of Poland-Lithuania, had been given an amber crucifix and altar goods.[32] Artful ambers were thus one field on which elite sibling rivalries sometimes played out.

Though many splendid pieces have now been lost to the ravages of time, the collection surviving in Florence is one of the finest in Europe.[33] It is so uniquely complete that developments in amber art and in individual practice can be tracked through it. Several signed and dated pieces show how quickly craftspeople in the Baltic were able to assimilate styles from further afield. Two altarpieces were made only five years apart and by the same man, Georg Schreiber, an amber-worker in Königsberg.[34] The first, for example, shows the influence of local northern art and architecture, particularly church furniture, but the second is clearly inspired by the new facade of St Peter's in Rome (illus. 81, 82). Georg Schreiber must have had access to prints showing the basilica, and indeed many ambers directly copy or adapt engravings. These too can be useful dating aids, although the fact that they could be used, reused, copied and reprinted certainly complicates this.

Technique can be a more reliable way to date objects. In the early 1600s, amber altarpieces were complex constructions of conjoined panels. This used amber's translucency to best effect but severely limited their size. When wooden substructures began to be used, altarpieces grew. Amber was applied directly atop the wood in a technique today called encrustation. This meant that light no longer passed through the amber, but craftspeople combatted this by facetting and engraving amber like gems and setting these cut pieces atop light-reflecting metal foils lacquered in a variety of colours. Using a wooden carcass also allowed for the incorporation of compartments. These are lined with mirror-glass and were suitable for relics or sacred figures. Some even incorporate captivating little amber scenes, such as the Last Supper – all in amber.

81, 82 Small amber altarpieces recorded as being in the chapel of Archduchess
Maria Maddalena of Austria, Grand Duchess of Tuscany, made by Georg Schreiber,
Königsberg, dated 1614 and 1619, respectively.

AMBER AND INGENUITY

Amber artisans were not only skilled, they were inventive, and ingenuity was often key to their works' appeal. Their ability to transform amber into interactive objects spoke to collectors who loved to be taken by surprise and filled with wonder. Fynes Moryson visited the Florentine collections in 1594. Later he wrote how impressed he had been. Alongside an amber cup and a lizard inclusion, he had seen a clock.[35] Contemporary inventories say nothing of a timepiece, but they do record an amber ship with crew. It is possible that this 'clock' was an automaton, for such amber ship automata were in other princely collections. In Berlin, there was even an amber automaton depicting miners at work. Though these were initially made in Europe, by the eighteenth century whimsical clockwork junks embellished with

83 Wine fountain, Königsberg, *c.* 1610, amber and gilt bronze.

84 Christoph Labhart, attrib., gondola-shaped vessel
depicting embracing lovers, Kassel, *c.* 1680/90.

amber were being imported back to Britain and Germany from China. The Florentine ducal collections were also home to a little artillery cannon of amber. Few such cannon survive today. Their adjustable barrels and movable carriages made them vulnerable to damage and they were clearly enjoyed. One later Dutch collector presented his as if on a bulwark, ready to barrage beholders. Spectacle was an essential element of presenting a collection in the early modern period, and amber was well suited to drama. Imagine wine tripping down the tiers of amber table fountains (illus. 83), looking through amber-lensed magnifying and eyeglasses, reading a book bound with it or even playing an amber flute.

There were risqué pleasures to be had too. In Milan, the cover of Manfredo Settala's pocket sundial hid 'a beautiful Flemish lady'. 'The stuff of dreams', she drew 'her essence from the candid amber' and radiated such loveliness that it was said that she could 'make you lose your heart'.[36] Tempting images of women were sometimes revealed when the lid of a tankard was flipped up or its contents – although it is not clear if they were actually capable of containing liquid – drained. It was claimed that drinking alcohol which had been in contact with amber caused excessive drunkenness, and these images of women hidden in drinking vessels speaks to Bacchic loss of inhibition. The sexual imagery on some cups is especially explicit and there are sensually sprawling women, sometimes even entangled lovers (illus. 84). Amber was also recommended to those who feared their partners had committed indiscretions and wished to force admissions.[37] If these indiscretions had invited unwanted side-effects, then venereal diseases could also be treated with powdered amber (sprinkled on a boiled egg or in wine).

VALUES

Cost and rarity, artistic skill, ingenuity, size, functionality, history, theatricality, even ideas about health and wellbeing: there were a host of reasons to treasure an object in amber. These might also be responsible for items being put to uses for which they were not initially intended. The duke of Mantua's amber figures of the Virgin Mary, Jesus Christ and twelve apostles, amber altarpiece, amber chalice and paten, amber candlesticks, amber cruets and amber pyx were not stored in his palace chapel or a sacristy but in the famous Hall of Troy, a reception space decorated by the celebrated painter Giulio Romano. In Florence, a set of apostles and a crucifixion even decorated the inside of a piece of furniture. Gleaming golden against the black ebony of the structure, the figures seared themselves onto John Evelyn's memory. In these cases, artistic and material value have been placed before devotional functionality.

Sacred ambers were even used for their monetary value. In the mid-twentieth century, curators in Florence found a note inside one of the Medici amber altarpieces in the Museo degli Argenti (now known as the Tesoro dei Granduchi). The note suggested that the altarpiece had been pawned in Danzig only a couple of years after completion.[38] Amber rosaries might also be used in bets. During the Papal Conclave of 1549–50, the Venetian ambassador reported seeing cardinals betting amber rosaries on the length of the assembly. During the same Conclave, Margaret of Austria presented the cardinal of Trent with a string of amber prayer beads and the thinly veiled message 'that he well knew how to make such a pope as would be confirmed by the Emperor'.[39]

Hard to come by for nearly everybody in this period, amber was a key material of ingratiation, and citizens of Prussia and the Polish-Lithuanian Commonwealth leveraged their advantage. Need to get the pope's attention? Then giving 'a heart of gold-gleaming amber, a half-finger in length ... carved [with] the image of John the Baptist as a child' to a cardinal might just do the trick.[40] Need some relics to increase your cathedral treasury? Why not send a handsome casket to a powerful prince. Need political support? Perhaps giving amber to those with a seat at the negotiating table might help. While representing the papacy in negotiations to end the Thirty Years War, Cardinal Francesco Barberini amassed a collection of five Virgin and Child figures and an apparently unique amber bust of Pope Urban VIII.

Insiders and outsiders alike saw Poland-Lithuania and Prussia as the cradle of amber. The association was taught in dry geography books and actively encouraged by their nobility and citizens. When the wife of the chancellor to the crown of Poland-Lithuania gave a prominent gift of amber to the famous shrine at Loreto in Italy this not only honoured the Church and demonstrated her piety, but the amber oil lamp, amber cruets for water and wine, amber altar candlesticks, amber basin, amber pax and amber-stemmed golden chalice were also a spectacular advert for her homeland. On a more personal level, while studying in Rome in 1640, one young man wrote to his father in Krakow asking him to supply an amber rosary or 'some other object made in amber' with which to fulfil a recent request from an acquaintance for 'something beautiful from Poland'.[41]

Giving amber as a way of stressing and strengthening the identification of material and place continues today. Mariusz Drapikowski and his Gdańsk-based studio specialize in the creation of sacred

Following the establishment of the British Museum in 1753, many important collections opened to the public, eventually becoming the museums still visited today. Magnificently worked ambers of Baltic origin were the pride of museums in Berlin, Dresden, Kassel and Munich, all formerly the residences of major rulers. Tragically, the exemplary Berlin collections, amassed by the rulers of Prussia, were destroyed in the Second World War. Austria and Russia are also home to significant collections. Ambers amassed by the Habsburg dynasty are split between Vienna and Innsbruck; Russian imperial ambers can be seen in Moscow and St Petersburg.[51] There are also notable collections in Denmark, and south of the Alps, collections in Florence, Modena and Naples stand out. In England, the Victoria and Albert Museum in London, founded on royal initiative but not on royal collections, has slowly built a collection to rival the old guard. Curators acquired the first European ambers in the mid-1850s, and there were further flurries in the 1870s and the 1920s, but the collection changed radically in 1950 with the generous gift of Walter Leo Hildburgh.

It is perhaps unsurprising that European museums primarily care for European ambers. Where there are collections of Asian ambers, these are typically small and concentrated on Tibetan and Burmese ethnographica, masterful carvings and snuff-bottles. The last in particular were popular among nineteenth-century Western collectors. Britain's historical interest in southern China, including its administration of Hong Kong, meant that Chinese amber arte-facts were available and collected in the United Kingdom long before interest developed in the United States. Auction records show that there was a resale market for Qing ambers in London from the 1830s onwards if not earlier.

COLLECTING IN THE LATER NINETEENTH CENTURY

In Europe, dealers and auctioneers were significant sources of historical ambers, and many collectors and new non-royal museums relied on them to develop their collections. In the nineteenth century there were still many museum-worthy ambers available to meet this demand. Amber cups, caskets and cabinets, figures – notably a rare set of the twelve apostles with Jesus Christ and a rarer St Elisabeth – chessboards and chess pieces, tankards, salvers, snuffboxes and cutlery were all sold in London auctions in the 1800s, having been owned by peers of the realm, politicians, poets and painters, judges, ministers, school masters and museum employees. There were even objects with royal provenance. Three notable examples are Queen Charlotte's amber chessboard sold in 1819; an amber cabinet reputedly made in the 1660s for the 'Princess of Bavaria' (or the 'Queen of Bohemia') sold in 1822 at the sale of the contents of William Beckford's Fonthill Splendens; and 'an antique jug with handle, formed of amber, engraved over and mounted in silver, formerly in the possession of the Empress Josephine' auctioned in 1849. North America's biggest collection of artistic ambers, today in the Museum of Fine Arts in Boston, was amassed by an individual, William Arnold Buffum, in the closing decades of the century.[52] It was also at this time that development in China, particularly the construction of the railway network, allowed previously remote regions and undisturbed sites of historical interest to be reached. The plundering of ancient tombs brought many early ambers onto the market, expanding the realm of acquirable ambers beyond those of the recent Qing dynasty.

COLLECTING IN THE TWENTIETH AND TWENTY-FIRST CENTURIES

Many American collectors and museums, often through collectors' bequests, were able to add Chinese ambers to their collections at the end of the nineteenth century. American interest in Chinese art, including ambers, is well illustrated by the sales in New York of artefacts owned by Chinese noblemen following the Chinese Revolution in 1911. The Revolution also precipitated the establishment of the Palace Museum in the Forbidden City, the collections of which are based on the Qing imperial collection and which include many fine pieces from that and earlier eras. Today the greatest collections of Chinese

88 Lucjan Myrta, *Self-Portrait in Amber,* 2009, amber and wood.

ambers outside of China are concentrated in North America, for example the Drummond Collection at the American Museum of Natural History, New York, and the Reif Collection in the Art Gallery of Great Victoria, Canada.

The history of artful ambers in the twentieth century is exceptionally complex and highly emotional. At the end of the Second World War, German refugees fleeing the Soviet invasion of East Prussia did as adventurers, explorers and travellers had done before them and carried precious family ambers to safety on their persons. Many later donated them to memorial museums founded where they came to settle in the new West Germany. The 1960s and '70s also saw the establishment of the amber collection in Ribnitz-Damgarten, Germany, and the founding of the amber museums at Palanga in Lithuania and Malbork in Poland. Fortunately for the latter, the sale of the Kitson Collection, primarily an Asian art collection, released both European and Chinese ambers onto the market. The sale coin-cided with the establishment of the museum, which can now claim to be the first public institution to attempt the presentation of ambers from East and West together. Their curator was also one of the first to open the collections to the work of contemporary jewellers and artists such as Lucjan Myrta (illus. 88). This has also been a focus of the Kaliningrad Amber Museum (established 1972), which displays new works alongside reconstructions of pieces in other collections and historical works gifted by the Kremlin. Even more recently, Gdańsk (formerly Danzig) established its own representative and world-class collection, including rare signed items and usually overlooked works in, for example, the Art Deco style.[53]

Private individuals can still buy small worked ambers from the early twentieth century, whether European or Asian, fairly easily at flea and antique markets or online, and there are a handful of special-ist dealers for historical pieces. There have been spectacular museum acquisitions in recent years, like the Rijksmuseum's purchase of an amber games-box believed to have been given to Anne, Princess Royal of Great Britain, and William IV, Prince of Orange, when they married in 1734. Going forward the waters look choppy. In 2018, the Ivory Act completely banned the trade in ivory in the United Kingdom. This ban includes sales of artistic cultural heritage where ivory amounts to more than 10 per cent of the object and affects many Baltic ambers, given the use of ivory for contrast and detail.

Does artistic amber have a future? Alatyr, the international competition staged biennially by the Amber Museum, Kaliningrad, is both a health-check and a call to arms. The event is accompanied by a conference addressing such critical issues as the place of amber in the education of contemporary jewellers and visual artists. Practical training in and experience of the material's unique characteristics are essential prerequisites if the art of amber is to have a future. These skills remain concentrated in Russia and Poland, where there are already long traditions of working. However, large busts of classical figures, Chinese pieces and sculptures of birds and reptiles made from Dominican, Mexican and Sumatran amber have recently been offered

Messer, were either spoliated, destroyed or travelled miles across the world in the luggage of their refugee owners (illus. 95). The Israel Museum in Jerusalem has a fine collection of ambers pertaining to the global Jewish diaspora as well as Jewish culture in Europe, and there is considerable scope for more work in this area.

AMBER FURNITURE

Apart from the Amber Room, the single greatest amber loss of the Second World War was the Prussian royal collection in Berlin. Without doubt the best collection of worked amber in Europe at any time, this had been rebuilt following the destruction of the first in the Thirty Years War. It encompassed many curious and playful pieces, including two farms, a 'military entrenchment' and 'equipment for a doll's kitchen', as well as musical instruments. While no other collection could boast the wealth and breadth of the one in Berlin, it was certainly not the only site of spectacular and unexpected types of objects. The rulers of Prussia gifted impressive pieces, particularly pieces of amber-clad furniture, to their peers across the continent – most of which were so delicate that they have now been lost to history.

One of the first types of furnishing to be attempted in Europe was the chandelier. The earliest account dates to the end of the six-teenth century.[14] It says nothing of the actual appearance of these chandeliers, and other inventory entries, like the description of Queen Anne's 'candlesticke of amber to hang upp in braunches in a double woodon box' among her possessions at Denmark House, London, in 1619, are equally scant. The same description was re-peated in 1651 when said 'candlesticke' was sold from among the effects of the ill-fated King Charles I.[15] Luckily, one chandelier was a major star. It was more than a metre high, weighed about 7 kilograms (15½ lb) and was in the collections of the grand dukes of Tuscany in Florence. The fantastical light fitting comprised 'three-tiers . . . with eight arms per tier, and ovals and roundels full of figures and histories in white amber and with an eagle on top'.[16] For the length of its existence, initially hanging from the mother-of-pearl-encrusted

96 Chandelier, Königsberg, 1660/70, amber.

cupola of the *Tribuna* and later in the light of a south-facing window in the *Gabinetto di Madama*, the chandelier attracted wonder and comment. Visitors were particularly curious about its origin. In the 1650s, the English traveller Richard Lassels was told it had been a gift from Duke Johann Georg of Saxony to Grand Duke Cosimo II's third son. It was said that Johann Georg had received it from Sophia, wife of Georg Friedrich of Brandenburg-Ansbach.[17] Nearly three-quarters of a century later, Johann Georg Keyssler was told that the figures were 'busts of princes and princesses of [the] illustrious house [of Brandenburg]' from which the chandelier had been a gift.[18] This agrees with what is now known about a payment for a counterweight in 1618 and other contemporary Florentine sources which name Grand Duke Cosimo rather than his son as the recipient. If these dates are correct, the donor is most likely to have been Johann Sigismund of Brandenburg, who married the daughter of Duke Albrecht Friedrich of Prussia and inherited the territory on his father-in-law's death in 1618.

Chandeliers in amber were certainly impressive, as later surviving examples show (illus. 96). They were not intended as practical light fixtures: the chandelier sent to King Frederik of Denmark in 1653, for example, was in response to a personal request for amber for his art collections.[19] Other potentates saw their potential as impressive gifts. The chandelier sent to Tsar Alexei Mikhailovich in Moscow in 1673 was initially hung up and admired but was, said the tsar to the representative who brought it, to be sent to Persia with the next embassy.[20] The Ottoman Empire and the Kingdom of Persia were courted by European rulers as gateways to the East and commodities such as silk. Elector Friedrich Wilhelm dispatched a second chandelier to Moscow in 1689.

Both Russian chandeliers are lost, but drawings of the latter were found and published before being destroyed themselves in the Second World War. They were prepared by its maker Michael Redlin and are a rare opportunity to read a craftsperson's own description of their work. His two-tiered, twelve-branched chandelier had been made from 'unadulterated amber in large pieces'. The light fitting was decorated with 'Roman and German emperors and heroes artfully painted on gilded foil and covered with clear amber' and took him more than two years to complete.[21] It was common for work to extend over several years. Amassing the right type of amber alone might take even longer. The amber-turner-turned-merchant Johann Koster needed five years to collect enough for the mirror frame, which he dedicated his final years to making.[22] Indeed, a single mirror frame might require more than 1,500 individual pieces of amber.[23] Given the substantial amounts of waste involved in working amber, this would have required perhaps as much as 15 kilograms (33 lb), the amount one craftsperson said they needed to make an amber-clad table. In terms of both material and skill, the value of these objects was immense. John Evelyn estimated the contemporary value of Queen Mary II's large amber cabinet and table mirror to be £4,000, a sum then capable of buying nearly 750 horses, and today equivalent to £480,000.[24]

Like chandeliers, the first mirror frames are recorded around 1600. The earliest were small, as dictated by the then-possible size of mirrored glass panels. Many were probably hand glasses supplied, as Anna of Denmark's was, with velvet-lined cases.[25] Others were wall-mounted (illus. 97).[26] Mirrors were luxury items, and mirrors framed in amber even more so. Like amber chandeliers, mirror-frames were royal gifts *par excellence.* One of the first grand commissions was, like the chandelier, initially a gift for the tsar but was presented to Louis XIV of France, the revered Sun King. Nicolaus Turow needed six months to customize the existing frame to add the arms of France, sunbursts and the king's motto.[27] Transport was held up since dry weather had reduced water levels in the rivers and canals. When it finally arrived in Paris, it was a marvellous sight. Louis is said to have commented on the skill demonstrated in the carvings, which showed scenes from Ovid's *Metamorphoses.* Like so many other objects, this frame is no longer traceable. It is believed to have been sent to Siam (Thailand) in 1687, alongside a further mirror frame, two amber caskets and a boat-shaped amber drinking vessel.[28] High value not only meant that gifts of mirrors were recycled, but that mirrors were

97 Mirror in the form of an epitaph, Königsberg,
late 16th century, amber, ivory and mirror glass.

98 Drawing showing the throne made of ivory and amber commissioned
by Friedrich Wilhelm I, c. 1677, paper and watercolour.

looted in times of conflict. The 'Venetian mirror in an amber frame'
in Karl Gustaf Wrangel's Skokloster Castle is reputedly booty from
the Thirty Years War.[29]

Such items must have been spectacular, but no chandelier or mirror has the renown of the great amber throne commissioned in 1676
to mark the upcoming twentieth anniversary of Emperor Leopold I's
coronation.[30] The electors of Brandenburg did not retain their own
dedicated amber carver, as the kings of Denmark, the landgraves of
Hessen and the electors of Saxony did. It is a remarkable omission

oily particles by showing that no chemist had been able to replicate the process. He noted, among other things, that amber's specific gravity (its density relative to water) was remarkably close to pine resin. Before long, Lomonosov's view that amber was a very old, very hard resin became widely accepted. It would take time for agreement on exactly how old to be reached. Today, of course, it is known that different ambers have different ages.

The consensus that amber was a resin, albeit an ancient one, may have combined with the greater availability and application of resins in day-to-day life to reduce its power as a material of luxury and result in a fall from fashion. In 1742, members of the Königsberg guild appealed to the king to employ more amber in his international diplomacy and to 'rule that your ambassadors, ministers and courtiers draw attention to our amber wares . . . and thereby facilitate their greater popularity'.[48] They argued that their financial situation was so strained as to prevent them acquiring the raw material of their trade, and it was soon decreed that the volume of amber required for the examination taken to become a master amber worker would be halved as apprentices could not afford basic raw materials. As traced elsewhere in this book, the industry continued, but as one which supplied different customers, needs and locations.

Across the globe, in China under the Qianlong emperor, stunningly wrought objects would soon reach the height of their popularity. While amber's fortune declined in Europe, in China its luxury status grew. So great was the esteem in which amber was held that it was even deployed (completely impractically given its fragility) in the production of *ruyi*. These sceptres – time-honoured symbols of authority and political power – were wielded in imperial ceremonies and given as high-profile gifts. One rare survivor in the Palace Museum Beijing is dated to 1770 by an inscription on the back. A further text explains that it was the personal gift of the Empress Dowager. It is a remarkable, delicate artefact which, at more than 35 centimetres (13¾ in.) long, must have required a substantial chunk of amber and tested the skills and material knowledge of its makers. The same artisans, many employed in the palace workshops, crafted the spectacular mountain landscapes, populated by intricately carved figures, that continue to enchant to this day (illus. 104).

CONCLUSION

This book has been an attempt to develop a history of human engagement with amber across at least three millennia and has touched upon all the world's continents but Antarctica. It spans a considerable time frame and paints pictures broadly. It has not been possible to address all the nuances or complexities of amber, not least because the word itself, in whichever language is used, has been a historically and culturally flexible label. Today amber continues to describe fossilized resins from a wide range of geological periods and geographical locations. A truly global, transnational history of amber is still waiting to be written.

Books dealing with the cultural and natural history of natural resources are, in their essence, not the obvious place to look for predictions concerning the future of these materials. Their focus is understandably their historical significance. However, as my experimental exploration shows, the historical interrogation of amber has only just begun. This is particularly true in respect of difficult histories, not least regarding the role raw and bead amber played in European colonialism and imperialism. Addressing such uncomfortable truths is key to ensuring amber and its study remains relevant now and in the future.

This book began with the question: what is amber? This has been asked for centuries by people who have encountered it. Today it is common knowledge that amber is a fossilized resin. With that mystery solved, future enquiries are more likely to focus on its whos, wheres, hows and whens. In order to ensure an economically sustainable future for amber, wherever it originates, it must begin to respond to the changing expectations of more environmentally conscious consumers. Transparent and ethical sourcing will be crucial. This will necessarily mean better methods of monitoring the global amber supply chain from mine to market. Going forward, consumers will certainly also seek to be reassured that miners are paid a fair price, work in safe conditions, are not underage, mine legally acquired land and use environmentally responsible and sustainable practices. In this era of climate crisis, it cannot be overlooked that unregulated mining is an environmental disaster, causing deforestation, erosion, loss of habitat and land degradation.

Everyone and everything has a role to play in slowing climate change, and amber is no exception. Even the simplest consumer measures can help. Lovers of amber jewellery, for example, may wish to consider purchasing pre-owned ambers or having antiques reworked using eco-friendly methods; they may also wish to ensure that the precious metal used in mounting is ethically sourced too. Surely there is no stronger message as to amber's potential for new discoveries than the fact that prominent palaeo-entomologists are arguing vehemently and publicly against suggestions that burmite should not be studied due to ethical and humanitarian concerns. They caution that the resulting loss of material and information will be a 'huge drawback for science in Myanmar and the rest of the world'.[49] Recent spectacular discoveries in amber have challenged established scientific fact. With the earth facing unprecedented future challenges, amber, the natural time capsule and preserver of key information about its evolutionary history, promises to offer invaluable insights into what might come next for humans and the planet.

References

one Amber: What, When, Where?

1 M. C. Bandy and J. A. Bandy, trans., *Georgius Agricola: De natura fossilium (Textbook of Mineralogy)* (New York, 1955), p. 71; Giuliano Bonfante, 'The Word for Amber in Baltic, Latin, Germanic, and Greek', *Journal of Baltic Studies*, XVI/3 (1985), pp. 316–19; and Faya Causey, 'Anbar, Amber, Bernstein, Jantar, Karabe', in *Bernstein, Sigmar Polke, Amber*, exh. cat., Michael Werner Gallery (New York, 2007).

2 This chapter draws on: Andrew Ross, *Amber: The Natural Time Capsule*, 2nd edn (London, 2010); Norbert Vávra, 'The Chemistry of Amber – Facts, Findings and Opinions', *Ann. Naturhist. Mus. Wien*, CXI/A (2009), pp. 445–74; Jorge A. Santiago-Blay and Joseph B. Lambert, 'Amber's Botanical Origins Revealed', *American Scientist*, XCV/2 (2007), pp. 150–57; Jean H. Langenheim, *Plant Resins: Chemistry, Evolution, Ecology, Ethnobotany* (Portland, OR, 2003); David A. Grimaldi, *Amber: Window to the Past* (New York, 1996); George Poinar and Roberta Poinar, *The Quest for Life in Amber* (Reading, MA, 1994); and Helen Fraquet, *Amber* (London, 1987).

3 George Poinar and Roberta Poinar, *The Amber Forest: A Reconstruction of a Vanished World* (Princeton, NJ, 1999).

4 Alexander P. Wolfe et al., 'A New Proposal Concerning the Origin of Baltic Amber', *Proceedings of the Royal Society of London B: Biological Sciences*, CCLXXVI (2009), pp. 3403–12. The seminal studies are: C. W. Beck, E. Wilbur and S. Meret, 'The Infrared Spectra of Amber and the Identification of Baltic Amber', *Archaeometry*, VIII (1965), pp. 96–109; and C. W. Beck at al., 'Infra-Red Spectra and the Origin of Amber', *Nature*, CCI (1964), pp. 256–7.

5 Vávra, 'The Chemistry of Amber'; F. Czechowski et al. 'Physicochemical Structural Characterisation of Amber Deposits in Poland', *Applied Geochemistry*, XI (1996), pp. 811–34.

Giacomo Fantuzzi: Diario del viaggio europeo (1652) (Warsaw and Rome, 1998); and Wojciech Tygielski, ed., *Giacomo Fantuzzi: Diariusz podróży po Europie (1652)* (Warsaw, 1990). See also Shackleton Bailey, trans., *Martial*, vol. I, bk IV, epi. 32.

51 James O'Brien, *The Scientific Sherlock Holmes: Cracking the Case with Science and Forensics* (Oxford and New York, 2013), p. 157.

52 Alexander Pope, *An Epistle from Mr Pope, to Dr Arbuthnot* (London, 1735), vol. I, pp. 9, 167–70.

53 Bacon, *Sylva*, p. 33.

54 Translation from Della Porta, *Natural Magick*, pp. 186–8.

55 Kundmann, *Rariora naturae*, pp. 219–26; and Johann Georg Keyssler, *Travels through Germany, Bohemia, Hungary, Switzerland, Italy and Lorrain*, 2nd edn (1757–8), vol. I, p. 432.

56 W.R.B. Crighton and Vincent Carrió, 'Photography of Amber Inclusions in the Collections of National Museums Scotland', *Scottish Journal of Geology*, XLIII/2 (2007), pp. 89–96.

57 David Penney et al., 'Extraction of Inclusions from (Sub)fossil Resins, with Description of a New Species of Stingless Bee in Quarternary Colombian Copal', *Paleontological Contributions*, VII (May 2013), pp. 1–6. For the most recent developments, see E.-M. Sadowski et al., 'Conservation, Preparation and Imaging of Diverse Ambers and Their Inclusions', *Earth-Science Reviews*, CCXX (2021), unpaginated.

six Accessorizing with Amber

1 J. B. Rives, trans., *Tacitus: Germania* (Oxford and New York, 1999), chap. XLV, ll. 4–6.

2 Robert Hellbeck, 'Die Staatliche Bernstein-Manufaktur als Trägerin der Preußischen Bernstein-Tradition', in *Preußische Staatsmanufakturen; Ausstellung der Preußischen Akademie der Künste zum 175jährigen Bestehen der Staatlichen Porzellan-Manufaktur* (Berlin, 1938), pp. 103–7.

3 See Rachel King, 'Bernstein. Ein deutscher Werkstoff?', in *Ding, Ding, Ting: Objets médiateurs de culture, espaces germanophone, néerlandophone et nordique*, ed. Kim Andringa et al. (Paris, 2016), pp. 101–20.

4 Ulf Erichson, ed., *Die Staatliche Bernstein-Manufaktur Königsberg: 1926–1945* (Ribnitz-Damgarten, 1998), p. 23. My translation.

5 Wilhelm Bölsche, 'Der deutsche Bernstein', *Velhagen und Klassings Monatshefte*, II (1934/5), pp. 89–90. My translation.

6 Erichson, ed., *Die Staatliche Bernstein-Manufaktur Königsberg*, and Hellbeck, 'Die Staatliche Bernstein-Manufaktur', pp. 105–7.

7 Alfred Rohde, *Das Buch vom Bernstein*, 2nd edn (Königsberg, 1941), p. 21.

8 'Bernstein als urdeutsche Schmuck', *Die Goldschmiedekunst*, XIX (1933), p. 433. My translation.

9 Alan Crawford, *C. R. Ashbee: Architect, Designer and Romantic Socialist*
   (New Haven, CT, and London, 1985), p. 350.

10 Rainer Slotta, 'Bernstein als besonderer Werkstoff', in *Bernstein. Tränen
   der Götter*, ed. M. Ganzelewski and R. Slotta, exh. cat., Deutsches Bergbau-
   Museum (Bochum, 1996), pp. 433–8.

11 Michael Ganzelewski, 'Bernstein – Ersatzstoffe und Imitationen', in *Bernstein.
   Tränen der Götter*, ed. Ganzelewski and Slotta, pp. 475–81, especially
   pp. 476–8.

12 Norbert Vávra, 'Bernstein und Bernsteinverarbeitung im Alten Wien', in
   Bernstein. Tränen der Götter, ed. Ganzelewski and Slotta, pp. 483–91.

13 Trade card for James Cox, goldsmith, at the Golden Urn, in Racquet Court,
   Fleet Street, London, British Museum, Sir Ambrose Heal collection of trade-
   cards, Heal, 67.99.

14 Eugen von Czihak, 'Der Bernstein als Stoff des Kunstgewerbes', in *Die
   Grenzboten. Zeitschrift für Politik, Literatur und Kunst* (Berlin and Leipzig,
   1899), pp. 179–89, 288–98.

15 Gesellschaft zur kunstgewerblichen Verwertung des Bernsteins GmbH.

16 Von Czihak, 'Der Bernstein als Stoff des Kunstgewerbes', pp. 288–9. My
   translation.

17 Georg Malkowsky, 'Das samländische Gold in Paris', in *Die Pariser
   Weltausstellung in Wort und Bild*, ed. Georg Malkowsky (Berlin, 1900),
   p. 138. My translation.

18 Otto Pelka, *Bernstein* (Berlin, 1920), p. 136. My translation.

19 Ibid.

20 Bettina Müller, 'Werke von Toni Koy, Goldschmiedin in Königsberg',
   www.ahnen-spuren.de, 19 September 2018; Jan Holschuh, Hans Werner
   Hegemann and Max Peter Maass, *Jan Holschuh – Bernstein, Elfenbein,
   Aluminium*, exh. cat., Deutsches Elfenbeinmuseum (Erbach, 1981);
   *Der Bildhauer Prof. Hermann Brachert: 1890–1972. Austellung zum 100.
   Geburtstag, Plastiken, Bernsteinarbeiten, Zeichnungen*, exh. cat.
   (Ravensburg, 1990).

21 'Simonis Grunovii, Monachi Ordinis Praedicatorum Tolkemitani Chronici',
   in P. J. Hartmann, *Succini Prussici physica et civilis historia* (Frankfurt, 1677),
   pp. 154–64, here pp. 156–7; Andreas Aurifaber, *Succini historia. Ein kurtzer:
   gründlicher Bericht woher der Agtstein oder Börnstein ursprünglich komme*
   (Königsberg, 1551), unpaginated. My translation.

22 Laurier Turgeon, 'French Beads in France and Northeastern North America
   during the Sixteenth Century', *Historical Archaeology*, XXXV/4 (2001),
   pp. 58–9, 61–82.

23 Dirk Syndram and Jochen Vötsch, eds, *Die kürfürstlich-sächsische
   Kunstkammer in Dresden* (Dresden, 2010), Das Inventar von 1587,
   ff. 232v/249v.

24 Turgeon, 'French Beads in France and Northeastern North America
   during the Sixteenth Century'.

25 Vatican City, Vatican Apostolic Archive, Miscellanea Armadio xv.80 (Itinerario di Iacomo Fantuzzi da Ravenna nel partire di Polonia dell 1652), ff. 25r–27v. Printed editions are: Piotr Salwa and Wojciech Tygielski, eds, *Giacomo Fantuzzi: Diario del viaggio europeo (1652)* (Warsaw and Rome, 1998); and Wojciech Tygielski, ed., *Giacomo Fantuzzi: Diariusz podróży po Europie (1652)* (Warsaw, 1990).

26 Wilhelm Tesdorpf, *Gewinnung, Verarbeitung und Handel des Bernsteins in Preußen von der Ordenszeit bis zur Gegenwart* (Jena, 1887), pp. 28, 109.

27 Ibid.

28 R. Schuppius, 'Das Gewerk der Bernsteindreher in Stolp', *Baltische Studien*, xxx (1928), pp. 105–99, here p. 114.

29 Taken from the statues of the Königsberg guild 1745, cited in Gisela Reineking von Bock, *Bernstein, das Gold der Ostsee* (Munich, 1981), p. 36.

30 Turgeon, 'French Beads in France and Northeastern North America during the Sixteenth Century'.

31 Eugene H. Byrne, 'Some Medieval Gems and Relative Values', *Speculum*, x (1935), pp. 177–87.

32 For example, M. Grazia Nico Ottaviani, ed., *La legislazione suntuaria, secoli xiii–xvi: Umbria* (Rome, 2005), p. 50.

33 M. C. Bandy and J. A. Bandy, trans., *Georgius Agricola: De natura fossilium (Textbook of Mineralogy)* (New York, 1955), p. 76.

34 Wilhelm Stieda, 'Lübische Bernsteindreher oder Paternostermacher', *Mittheilungen des Vereins für Lübeckische Geschichte und Alterthumskunde*, ii/7 (1885), pp. 97–112, here p. 109.

35 Jemima Kelly, 'Amber Growth Turns Red as Oversupply Knocks Value', www.ft.com, 30 May 2019.

36 Adrian Christ, *The Baltic Amber Trade, c. 1500–1800: The Effects and Ramifications of a Global Counterflow Commodity*, MA thesis, University of Alberta, 2018, pp. 50–57.

37 An excellent recent study is Moritz Jäger, 'Mit Bildern beten: Bildrosenkränze, Wundenringe, Stundengebetsanhänger (1413–1600). Andachtsschmuck im Kontext spätmittelalterlicher und frühneuzeitlicher Frömmigkeit', PhD thesis, Geissen University, 2014.

38 *The Four Epistles of A. G. Busbequius, concerning His Embassy into Turkey* (London, 1694), p. 211.

39 Robert de Berquen, *Les Merveilles des Indes orientales et occidentales; ou, Nouveau traitte des pierres precieuses et perles* (Paris, 1669), p. 98; Ulisse Aldrovandi, *Musaeum metallicum in libros iv* (Bologna, 1648), p. 416.

40 Olga Pinto, ed., *Viaggi di C. Federici e G. Balbi alle Indie orientali* (Rome, 1962), pp. 97, 118–19; Pietro Della Valle, *Viaggi descritti in 54 lettere familiari, in tre parti, cioé la Turchia, la Persia e l'India* (Rome, 1650), vol. i, p. 737.

41 Pierre Pomet, *Histoire générale des drogues, traitant des plantes, des animaux, et des mineraux* (Paris, 1694), p. 84. My translation.

42 Robert Howe Gould, trans., *Théophile Gautier: Constantinople of To-Day* (London, 1854), pp. 117–18.

43 Ronald Gobiet, ed., *Der Briefwechsel zwischen Philipp Hainhofer und Herzog August d. J. von Braunschweig-Lüneburg* (Munich, 1984), p. 289. My translation.

44 Aurifaber, *Succini historia*; Anselmus de Boodt, *Le Parfaict joaillier; ou, Histoire des pierreries* (Lyons, 1644), pp. 410–29, especially pp. 418–22.

45 John Houghton, *A Collection for the Improvement of Husbandry and Trade* (London, 1727), vol. II, p. 65. See also D. E. Eichholz, trans., *Pliny: Natural History*, Loeb Classical Library (London and Cambridge, MA, 1962), vol. x, bk XXXVII, chap. XXII, l. 51, on amber and childhood illnesses. Farah Abdulsatar et al., 'Teething Necklaces and Bracelets Pose Significant Danger to Infants and Toddlers', *Paediatrics and Child Health*, XXIV/2 (May 2019), pp. 132–3.

46 Lu Liu, 'An Illustrated Manual for Regulating the Qing Society: A Discussion of Several Issues Relating to "Huangchao liqi tushi"', *Palace Museum Journal*, IV (2004), pp. 130–44. See also Xiaodong Xu, 中國古代琥珀藝術 (*Zhongguo gu dai hu po yi shu/Chinese Ancient Amber Art*) (Beijing, 2011); and Barry Till, *Soul of the Tiger: Chinese Amber Carvings from the Reif Collection* (Victoria, BC, 1999), p. 19.

47 English taken from *The History of That Great and Renowned Monarchy of China . . . lately written in Italian by F. Alvarez Semedo* (London, 1655), p. 9.

48 Xiaodong Xu, *Chinese Ancient Amber Art*, pp. 61–70.

49 John Cordy Jeaffreson, ed., *Middlesex County Records* (London, 1886), vol. I, bill dated 10 April 1583.

50 John M. Riddle, 'Amber: An Historical-Etymological Problem', in *Laudatores temporis acti: Studies in Memory of Wallace Everett Caldwell*, ed. Gyles Mary Francis and Davis Eugene Wood (Chapel Hill, NC, 1964), pp. 110–20.

51 Bandy and Bandy, *De natura fossilium*, p. 76.

52 German (*Bernstein*) Dutch (*barnsteen*), Polish (*bursztyn*), Hungarian (*borostyn*) and Swedish (*bärnsten*).

53 Joseph Browne, *A Practical Treatise of the Plague, and All Pestilential Infections That Have Happen'd in This Island for the Last Century*, 2nd edn (London, 1720), p. 50.

54 Berlin, Geheimes Staatsarchiv Preussischer Kulturbesitz, XX Hauptabteilung, Etatsministerium 16a 5, 'Abhandlung von Gregor Duncker über den Ursprung des Bernsteins als Arznei' (Treatise by Gregor Duncker on the Beginnings of Amber as a Medicine, c. 1538), ff. 2r–4v.

55 Giuseppe Donzelli, *Teatro farmaceutico, dogmatico, e spagirico*, 3rd edn (Rome, 1677), pp. 399–400. My translation.

56 Aurifaber, *Succini historia*, unpaginated; Johann Wigand, *Vera historia de succino Borussica* (Jena, 1590); and John Bate, *The Mysteries of Nature and Art: In Four Severall Parts*, 3rd edn (London, 1654), p. 217.

Penney, David, ed., *Biodiversity of Fossils in Amber from the Major World Deposits* (Manchester, 2010)

Poinar, George, and Roberta Poinar, *The Amber Forest: A Reconstruction of a Vanished World* (Princeton, NJ, 1999)

—, *The Quest for Life in Amber* (Reading, MA, 1994)

—, *What Bugged the Dinosaurs? Insects, Disease and Death in the Cretaceous* (Princeton, NJ, 2009)

Reineking von Bock, Gisela, *Bernstein, das Gold der Ostsee* (Munich, 1981)

Riddle, John M., 'Amber: An Historical-Etymological Problem', in *Laudatores temporis acti: Studies in Memory of Wallace Everett Caldwell*, ed. Gyles Mary Francis and Davis Eugene Wood (Chapel Hill, NC, 1964), pp. 110–20

—, 'Amber and Ambergris in Materia Medica during Antiquity and the Middle Ages', PhD thesis, University of Carolina, 1965

—, 'Pomum ambrae: Amber and Ambergris in Plague Remedies', *Sudhoffs Archiv für Geschichte der Medizin und der Naturwissenschaften*, XLVIII/2 (1964), pp. 111–22

Rippa, Alessandro, and Yi Yang, 'The Amber Road: Cross-Border Trade and the Regulation of the Burmite Market in Tengchong, Yunnan', *TRANS: Trans-Regional and -National Studies of Southeast Asia*, V/2 (2017), pp. 243–67

Rohde, Alfred, *Bernstein: Ein deutscher Werkstoff. Seine künstlerische Verarbeitung vom Mittelalter bis zum 18. Jahrhundert* (Berlin, 1937)

Ross, Andrew, *Amber: The Natural Time Capsule*, 2nd edn (London, 2010)

—, and Alison Sheridan, *Amazing Amber* (Edinburgh, 2013)

Santiago-Blay, Jorge A., and Joseph B. Lambert, 'Amber's Botanical Origins Revealed', *American Scientist*, XCV/2 (2007), pp. 150–57

Scott-Clark, C., and A. Levy, *The Amber Room: The Fate of the World's Greatest Lost Treasure* (New York, 2004)

Seipel, Wilfried, ed., *Bernstein für Thron und Altar: das Gold des Meeres in fürstlichen Kunst- und Schatzkammern*, exh. cat., Kunsthistorisches Museum, Vienna (Milan, 2005)

Serpico, Margaret, 'Resins, Amber and Bitumen', in *Ancient Egyptian Materials and Technology*, ed. P. T. Nicholson and I. Shaw (Cambridge, 2000), pp. 430–74

Smirnov, R., and E. Petrova, trans., *The Baltic Amber from the Collection in the State Hermitage Museum* (St Petersburg, 2007)

So, Jenny F., 'Scented Trails: Amber as Aromatic in Medieval China', *Journal of the Royal Asiatic Society*, XXIII/1 (2013), pp. 85–101

Sun, Zhixin Jason, 'Carved Ambers in the Collection of the Metropolitan Museum of Art', *Arts of Asia*, XLIX/2 (2019), pp. 70–77

Till, Barry, *Soul of the Tiger: Chinese Amber Carvings from the Reif Collection* (Victoria, BC, 1999)

Trusted, Marjorie, *Catalogue of European Ambers in the Victoria and Albert Museum* (London, 1985)

Vávra, Norbert, 'The Chemistry of Amber – Facts, Findings and Opinions', *Ann. Naturhist. Mus. Wien*, CXI (2009), pp. 445–74

Veil, Stephen, et al., 'A 14,000-Year-Old Amber Elk and the Origins of Northern European Art', *Antiquity*, LXXXIII/333 (2012), pp. 660–73

Volchetskaya, T. S., H. M. Malevski and N. A. Rener, 'The Amber Industry: Development, Challenges and Combating Amber Trafficking in the Baltic Region', *Baltic Region*, IX/4 (2017), pp. 87–96

Xiaodong Xu, 中國古代琥珀藝術 (*Zhongguo gu dai hu po yi shu/Chinese Ancient Amber Art*) (Beijing, 2011)

Zherikhin, V. V., and A. Ross, 'A Review of the History, Geology and Age of Burmese Amber (Burmite)', *Bulletin of the Natural History Museum, London (Geology)*, LVI (2000), pp. 3–10

Acknowledgements

I would like to thank Michael Leaman and Reaktion Books for commissioning *Amber* and for being understanding about the challenges I have faced while writing – not least two major museum redevelopments, several international house-moves, a growing family and a pandemic. Alex Ciobanu, Phoebe Colley and other members of the team have been generous with their time and expertise. My sincere gratitude to Dr Jill Cook and the anonymous readers of this book for their generous and insightful comments. All mistakes are of course my own.

I am deeply grateful to and miss Suzanne B. Butters, who shaped my approach to the subject immeasurably but who sadly died before this book was finished. I am also indebted to Tom Rasmussen, David O'Conner and Luca Mola, all of whom saw this research in its initial stages. I was encouraged to work in this field by Donal Cooper and Anne Matchette, and I have been supported and encouraged down the years by more people than I can name. I am enormously grateful to Iris Bauermeister, Cristina Cappellari, Faya Causey, Benoît Chauvin, Jill Cook, Spyros Dendrinos, Christopher Duffin, Godfrey Evans, Sarah Faulks, Alexandra Green, J. D. Hill, J. L. and S. J. King, Aleksandra Lipińska, Irina Polyakova, Ewa Rachoń, James Robinson, Andrew Ross, Judy Rudoe, Susan Russell, Anna Sobecka, Eva Stamoulou, Holly Trusted, Julia Weber and Erik Wegerhoff. I would also like to thank colleagues and peers at museums in Berlin, Boston, Edinburgh, Florence, Gdańsk, Glasgow, Kassel, Kaliningrad, London, Manchester, Munich, New York and Rome, and the staff of libraries and archives in all the same cities. Xiaodong Xu generously helped during the pandemic by sending scans of her work. I have had the opportunity to test my ideas at conferences, in seminars and in workshops. Some presentations went on to be published and I owe a debt of gratitude to all who have listened, commented, corrected and helped develop my thoughts. More than fifteen years of thinking about amber lie behind this book.

An enormous thank you to my talented colleague Craig Williams for the production of the illustrations for this publication. I am also grateful to the many museum, auction house, university and image library professionals who have assisted me with my requests for images and image licences. I would like to recognise those who shared images from their personal collections, provided advice and made connections with people on my behalf, foremost Jörn Barfod, Alex Ciobanu, Sarah Davis, Richard Evershed, Rosanna Falabella, Zuzana Francová, Yale Goldman, Stephen Hanson, Herman Hermsen, Michał Kosior, Alistair Mackie, Carlos Odriozola, Peter Pfaelzner, Irina Polyakova, Jan Rogalo, Iona Shepherd, Małgosia Siudak, Anna Sobecka, Elena Strukova, Lore Troalen, and Astrid Ubbink.

The research for this book enjoyed and was enriched by the generous support of the Arts and Humanities Research Council, the Studienstiftung des Abgeordnetenhauses Berlin, the British School at Rome, the Society for Renaissance Studies and the British Museum Scholarly Publication Fund.

Finally, I'd like to thank my family and friends for their forbearance.

Photo Acknowledgements

The author and publishers wish to express their thanks to the below sources of illustrative material and/or permission to reproduce it:

Alamy Stock Photo: 16 (incamerastock), 30 and 31 (Artokoloro), 69 (Rupert Sagar-Musgrave), 78 (Arterra Picture Library/Marica van der Meer), 85 (Karsten Eggert), 91 (dpa Picture Alliance Archive); © ArtSzok Tomasz Pisanko, Gdańsk: 74; © Ming Bai, Chinese Academy of Sciences (CAS), Beijing: 7, 53, 54; Bayerisches Nationalmuseum, Munich, photos © Bayerisches Nationalmuseum: 15 (Inv.-Nr. R 2757; photo Marianne Stöckmann), 75 (Inv.-Nr. MA 2478); photo Muriel Bendel (CC BY-SA 4.0): 11; photo Francesco Bini/Sailko (CC BY 3.0): 14; © Bodleian Libraries, University of Oxford (CC BY-NC 4.0): 102 (MS Arch. Selden. A. 1, fol. 47r); photo © bpk: 57; from Georg Braun and Franz Hogenberg, *Civitates orbis terrarum*, vol. II (Cologne, 1575), photo © The National Library of Israel, Jerusalem (The Eran Laor Cartographic Collection, Shapell Family Digitization Project and The Hebrew University of Jerusalem, Department of Geography – Historic Cities Research Project): 39; Burgerbibliothek, Bern, photo Codices Electronici AG/www.e-codices.ch (CC BY-NC 4.0): 34 (Mss.h.h.I.1, p. 304); from Johann Amos Comenius, *Orbis sensualium pictus*, part II (Nuremberg, 1754), photo Staatsbibliothek zu Berlin – Preußischer Kulturbesitz: 37 (B XVI, 7 R; http://resolver.staatsbibliothek-berlin.de/SBB00019B6700000000); from H. Conwentz, *Monographie der baltischen Bernsteinbäume: vergleichende Untersuchungen über die Vegetationsorgane und Blüten . . .* (Gdańsk, 1890), photo Universitäts- und Landesbibliothek Münster: 10 (RB 245); © Cordy's Auctions, Auckland, New Zealand: 103; courtesy Rosanna Falabella: 70; photo Alex 'Florstein' Fedorov (CC BY-SA 4.0): 93; Geheimes Staatsarchiv Preußischer Kulturbesitz, Berlin-Dahlem: 36 (XX. HA, Etatsministerium 16a 15, fol. 46); Geowissenschaftliche Museum der Universität Göttingen, photo © GZG Museum/G. Hundertmark: 23 (GZG.BST.10002 [old no. 58-002]); from Conrad Gessner, *De rerum fossilium, lapidum et gemmarum maximè, figuris & similitudinibus liber* (Zürich, 1565), photo

Zentralbibliothek, Zürich (FF 1264; https://doi.org/10.3931/e-rara-4176): 49; photo © Yale Goldman: 41; Grünes Gewölbe, Staatliche Kunstsammlungen, Dresden, photo © bpk/Staatliche Kunstsammlungen Dresden/Jürgen Karpinski: 67 (III 88 II/1); from Christoph Hartknoch, *Alt- und Neues Preussen Oder Preussischer Historien Zwey Theile, in derer erstem von desz Landes vorjähriger Gelegenheit und Nahmen . . .* (Frankfurt and Leipzig, 1684), photo Elbląska Biblioteka Cyfrowa, Elbląg (Pol.7.III.69): 9; from Philipp Jacob Hartmann, *Succini Prussici Physica & civilis historia: cum demonstratione ex autopsia & intimiori rerum experientia deducta* (Frankfurt, 1677), photos Zentralbibliothek, Zürich (NG 1909; https://doi.org/10.3931/e-rara-30618): 35, 40; from Daniel Hermann, *De Rana et Lacerta* (Krakow, 1583), photo Sächsische Landesbibliothek – Staats- und Universitätsbibliothek (SLUB), Dresden (Lit.Lat.rec.A.380, misc. 23; http://digital.slub-dresden.de/id428250165): 52; courtesy Heritage Auctions, HA.com: 73; © Herman Hermsen: 56; © Hofer Antikschmuck, Berlin: 55; The Israel Museum, Jerusalem, photo © The Israel Museum/Yair Hovav: 95 (received through JRSO (Jewish Restitution Successor Organisation), Wiesbaden collecting point number 5428, B50.02.1315 149/074); The J. Paul Getty Museum, Los Angeles: 90; photo © Kaliningrad Regional Amber Museum: 46 (КМЯ 1 No. 4210); Det Kongelige Bibliotek, Copenhagen: 19 (GKS 1633 4°, fol. 6r); © Michał Kosior, Amber Experts, Gdańsk: 1; Kunsthalle zu Kiel, photo © Kunsthalle zu Kiel/Sönke Ehlert: 62 (Inv.-Nr. 239); Kunsthistorisches Museum, Vienna, photo © KHM-Museumsverband: 77 (Schatzkammer, Kap 274); Kupferstichkabinett, Staatliche Museen zu Berlin, photo © bpk/Kupferstichkabinett, SMB/Jörg P. Anders: 98 (Inv.-Nr. 3135); courtesy Landesamt für Denkmalpflege im Regierungspräsidium Stuttgart: 27; Library of Congress, Prints and Photographs Division, Washington, DC: 94; © Alastair Mackie, courtesy of RS&A and All Visual Arts, London: 89; photo Ludmila Maslova (CC BY-SA 4.0): 43; The Metropolitan Museum of Art, New York: 26, 59, 104; Musée du Louvre, Paris, photo © RMN-Grand Palais (musée du Louvre)/ Daniel Arnaudet: 66 (OA7071); Museum für Naturkunde, Berlin, photos © bpk/ Museum für Naturkunde/Carola Radke: 51, 101; Museum Het Valkhof, Nijmegen, photo © Museum Het Valkhof: 20 (PDB.1988.7.ULP.1982.411.la); Museum of Fine Arts, Boston, photos © 2022 Museum of Fine Arts, Boston: 63 (bequest of William Arnold Buffum, 02.224), 71 (bequest of William Arnold Buffum, 02.91); Museumslandschaft Hessen Kassel, photos © Museumslandschaft Hessen Kassel: 84 (Inv.-Nr. KP B VI/I.53), 97 (Inv.-Nr. KP B VI/I.14); Muzeum Bursztynu – Muzeum Gdańska, photo © Muzeum Gdańska/M. Jabłoński: 48 (acquired with the support of the Kronenberg Foundation Citi Handlowy, MHMG/B/108); Múzeum mesta Bratislavy, photo © MMB/Ľudmila Mišurová: 76 (Inv. No. F-355); Nationalmuseet, Copenhagen (CC BY-SA 4.0), photos Roberto Fortuna and Kira Ursem: 21 (A48088), 25 (B1482); Nationalmuseet, Copenhagen (CC BY-SA 4.0), photo Arnold Mikkelsen: 22 (A54499); © National Museums Scotland: 32 (X.FC.8); © Carlos P. Odriozola and José Ángel Garrido Cordero: 44; © Die Österreichische Bernsteinstraße: 29; from Johann Posthius, *Germershemii Tetrasticha in Ovidii Metam. lib. XV . . .* (Frankfurt, 1563), photos Getty Research Institute, Los Angeles: 12, 13; photo S. Rae (CC BY